I0063882

MONOGRAPHS ON PHYSICS

EDITED BY

Sir J. J. THOMSON, O.M., F.R.S.

MASTER OF TRINITY COLLEGE, CAMBRIDGE

AND

FRANK HORTON, Sc.D.

PROFESSOR OF PHYSICS IN THE UNIVERSITY OF LONDON

MONOGRAPHS ON PHYSICS.

Edited by SIR J. J. THOMSON, O.M., F.R.S.,
Master of Trinity College, Cambridge;
and FRANK HORTON, Sc.D.,
Professor of Physics in the University of London.

8vo.

LONGMANS, GREEN AND CO.
39 PATERNOSTER ROW, LONDON
NEW YORK, BOMBAY, CALCUTTA, AND MADRAS

THE

EMISSION OF ELECTRICITY
FROM HOT BODIES

BY

O. W. RICHARDSON, F.R.S.

WHEATSTONE PROFESSOR OF PHYSICS, KING'S COLLEGE, LONDON

WITH DIAGRAMS

SECOND EDITION

LONGMANS, GREEN AND CO.

39 PATERNOSTER ROW, LONDON

FOURTH AVENUE & 30TH STREET, NEW YORK

BOMBAY, CALCUTTA, AND MADRAS

1921

PREFACE TO THE SECOND EDITION.

IN this edition I have extended certain sections of the book and re-written others, so as to bring it up to date. I have also taken the opportunity to correct a number of errors which appeared in the First Edition. The enormously rapid progress made in the technical developments of the subject in the last few years has made it more impossible than ever to deal with that aspect of the subject-matter in a book of this size.

O. W. R.

August, 1921.

PREFACE.

It will be seen from the following pages that the subject of the emission of electricity from hot bodies is one which has made rapid strides in recent years. It will also be clear that this field of inquiry still suggests for investigation many interesting questions which are either of theoretical or of practical importance. In dealing with the theory of the emission of electrons, one feels continually handicapped by the absence of a satisfactory and comprehensive theory of conduction for conductors of the metallic type. For this reason I have tried to make the treatment of this part of the subject as general as possible, and to reduce the part played by special theories to a minimum. Another difficulty lies in the interpretation of the facts bearing on the true nature of the contact potential difference between metals. In spite of a conflict lasting over a century, there still seems to be much room for difference of opinion here. This question is of fundamental importance in the interpretation of the theory of the emission of electricity from hot bodies.

It has seemed undesirable to include in the book an account of the numerous and important technical developments of the subject. Readers who are interested in these may, however, find useful the following list of references, arranged according to subjects:—*Wireless Telegraphy and Telephony*: Fleming, "Instrument for converting alternating currents into continuous currents," British Patent, No. 24,850 of 1904; De Forest, "The audion detector and amplifier," "Electrician," Vol. LXXII, p. 285 (1913); Reisz, "A new method of magnifying electric currents," *ibid.*, Vol. LXXII, p. 726 (1914); Langmuir, "The pure electron discharge and its applications in radio-telegraphy and telephony," *ibid.*, Vol. LXXV, p. 240 (1915); Armstrong, "Some recent developments in the audion receiver," *ibid.*, Vol. LXXVI, p. 798 (1916). *Production of X-Rays:* Coolidge, "A powerful Roentgen-ray tube with a pure electron discharge," "Phys. Rev.," Vol. II, p. 409 (1913). *Rectification of Alternating Currents:* Langmuir, loc. cit.; Hull, "A powerful source of constant high potential," "Phys. Rev.," Vol. VII, p. 405 (1916). *The Electric Arc:* MacKay and Ferguson, "Arcs in gases between non-vaporizing electrodes," *ibid.*, Vol. VII, p. 410 (1916).

In the last chapter I have included a brief account of the results of some experiments I have recently made on the electrons liberated by chemical action. Part of the cost of this investigation has been defrayed by a Government grant through the Royal Society.

For permission to publish certain of the figures I am indebted to the Royal Society, to the Cambridge Philosophical Society, to the American Physical Society, and to the Publishers of the "Philosophical Magazine".

Finally, I wish to express my thanks to Professor Newall for information bearing on the question of solar electricity considered on page 47, and to my wife and to Professor Horton for assistance with the proofs.

O. W. RICHARDSON.

KING'S COLLEGE, LONDON,
1 *May*, 1916.

CONTENTS.

CHAPTER I.

NATURE OF THE PHENOMENA.

IT is not intended in this book to give an account of all the electrical properties of bodies which depend upon temperature. In fact, the scope of the book is almost restricted to those phenomena which I have ventured to describe by the term thermionic. As is well known, all substances become conductors of electricity at sufficiently high temperatures. Not only is this the case, but solid and liquid substances have the power of conferring the property of electrical conductivity on the space which surrounds them. In other words, a charge of electricity tends to leak away from the surfaces of bodies at high temperatures. In general this happens in a vacuum as well as when the bodies are surrounded by a gaseous atmosphere. The study of these thermionic effects has led to many results of an interesting character, as we shall see. In practice it is often wellnigh impossible to separate the purely thermal effects from those caused indirectly by other actions which are conditioned by temperature. In this category effects due to chemical action are conspicuous. Chiefly for this reason I have added a chapter on Ionization by Chemical Action. At the same time I have omitted to describe the interesting electrical properties of flames, a subject which might perhaps have been expected to fall within the scope of the book. Those who are interested in flames will find an excellent account of their electrical properties in a recent work by H. A. Wilson.[1]

---

[1] "Electrical Properties of Flames and of Incandescent Solids," by H. A. Wilson (London University Press, 1912).

EARLY EXPERIMENTS.

The subject under consideration is not entirely of recent origin. In fact, it has been known for nearly 200 years that air in the neighbourhood of hot solids has the power of conducting electricity. Experiments on the subject were made by a number of physicists of the seventeenth century, including Du Fay,[1] Du Tour,[2] Watson,[3] Canton,[4] Priestley,[5] and Cavallo.[6] The phenomena appear to have attracted little further attention until the middle of the nineteenth century, when Becquerel[7] showed that air at a white heat was unable to insulate under a potential difference of a few volts. Somewhat later Blondlot[8] showed that the same was true even with a potential difference of 0·001 volt; he also found that the currents did not obey Ohm's law. An important discovery was made by Guthrie,[9] who showed that a red-hot iron ball in air could retain a negative charge but could not retain a positive charge. At higher temperatures this difference disappeared, electrifications of either sign being conducted away rapidly. This difference in the character of the discharge, according to the sign of the electrification, is sometimes described by the term unipolar and is of fundamental importance.

A systematic investigation of the electrical effects produced by incandescent solids was begun by Elster and Geitel[10] about 1880. Their method consisted in heating various metal wires by means of an electric current and examining the potential acquired by a neighbouring electrode under different circumstances. The hot wire was as a rule connected to the earthed

[1] " Mémoires de l'Acad." (1733).
[2] " Mém. de Math. et de Physique," XI, p. 246 (1755).
[3] " Phil. Trans." abridge. Vol. X, p. 296 (1746).
[4] *Ibid.*, Vol. LII, p. 457 (1762).
[5] " History of Electricity," p. 579.
[6] " Treatise on Electricity," Vol. I, p. 324.
[7] " Ann. de Chimie et de Physique," iii. Vol. XXXIX, p. 355 (1853).
[8] " C. R.," Vol. XCII, p. 870 (1881); Vol. CIV, p. 283 (1887).
[9] " Phil. Mag.," iv. Vol. XLVI, p. 257 (1873).
[10] " Ann. der Phys.," Vol. XVI, p. 193 (1882); Vol. XIX, p. 588 (1883); Vol. XXII, p. 123 (1884); Vol. XXVI, p. 1 (1885); Vol. XXXI, p. 109 (1887); Vol. XXXVII, p. 315 (1889); " Wien. Ber.," Vol. XCVII, p. 1175 (1889).

pair of quadrants of an electrometer, the other pair being connected to the electrode. Let us suppose that the wire is
maintained at a constant potential, and that all the quadrants
of the electrometer are connected together initially. In
general, an electric current is then flowing either from the
hot wire to the electrode or vice versa, and when the quadrants
are separated this current will give rise to a deflection of the
electrometer. The deflection will not increase indefinitely,
however, since the charging up of the electrode gives rise to
a back electromotive force which tends to stop the current.
Ultimately a limiting potential is reached which is sufficient
either completely to stop the current or to stop so much of it
that the rest just makes up for any small losses which may
arise from faulty insulation. The determination of this limiting potential under a great variety of conditions was the chief
object of most of Elster and Geitel's experiments. They
found that the magnitude and sign of the limiting potential
varied greatly in different circumstances. With a platinum
wire in air at atmospheric pressure this potential was positive
at low temperatures and increased in magnitude as the temperature was raised to a red heat, when a maximum value
was reached. After passing this point the potential fell almost to zero at a white heat. At lower pressures the results
were similar, except that the limiting potential, after passing
the temperature at which it reached zero, was found to change
sign and to acquire progressively increasing negative values as
higher temperatures were reached. The wires thus behaved
as though they had a tendency to give off positive electricity
at low temperatures and negative at high temperatures. At
some intermediate temperature equal amounts of each sign
would be given off; so that the potential acquired by the
electrode would be the same as that of the hot wire. The
temperature at which the change from positive to negative
took place was higher the higher the pressure of the air, and
it was also higher for new wires than for wires which had
been heated for a long time. It depended also on the nature
of the gas and on the material of the wire. With platinum
wires the phenomena in water vapour and the vapours of

sulphur and phosphorus were similar to those in air, but in hydrogen the electrode acquired a negative charge at all pressures up to and including atmospheric. With a copper wire in hydrogen, on the other hand, the electrode received a positive charge except when the pressure was quite low. Carbon filaments apparently gave rise to negative potentials under all circumstances.

Branly [1] used a method which is in some ways the opposite of that of Elster and Geitel. He measured the rate of *leakage* of electricity from an insulated conductor when placed in the neighbourhood of a hot body. In this way he obtained results in confirmation of those given by Elster and Geitel for platinum. He also found that the oxides of lead, aluminium, and bismuth, exhibited the opposite behaviour to that of various metals which had been tested; since in air at a red heat they lost a negative charge but not a positive charge.

An effect which occurs in electric lamps and was first observed by Edison is related to these phenomena. If an independent electrode is mounted in an incandescent lamp and arranged so that it can be connected through a galvanometer to either of the outside terminals of the lamp, a current is found to flow through the galvanometer when the connexion is made to the positive terminal but not when it is made to the negative terminal. A large number of experiments bearing on the question were made by Preece [2] and Fleming.[3] Fleming showed that the effects could be explained on the view that there was a vigorous emission of electricity from the negative end of a carbon filament even in the best possible vacuum. This conclusion was also in agreement with the earlier observations made by Elster and Geitel in their experiments on carbon filaments.

The Theory of Ions.

During the period which has just been under consideration the development of the subject was seriously handicapped by

---

[1] "C. R.," Vol. CXIV, p. 1531 (1892).
[2] "Roy. Soc. Proc.," Vol. XXXVIII, p. 219 (1885).
[3] *Ibid.*, Vol. XLVII, p. 118 (1890); "Phil. Mag.," Vol. XLII, p. 52 (1896).

the absence of any satisfactory theory to indicate the important lines of experimental investigation. This want was partially met, at the close of the nineteenth century, by the hypothesis which attributed the conduction of electricity by gases to the motion, under the influence of the electric field, of minute electrically charged particles or *ions*. Stimulated by the discovery of the Roentgen and Becquerel rays this hypothesis in the hands of Sir J. J. Thomson rapidly developed into a coherent theory capable of embracing all the known facts of gaseous discharges and of predicting many new phenomena hitherto unsuspected. Those who had studied the question felt that there was a definite connexion between the phenomena exhibited by gases when ionized, or made to conduct, under the influence of the Roentgen rays and other agencies, on the one hand, and the effects described in the last section on the other. In fact, the view of those effects which seems to have received most support at this time was somewhat as follows : It was supposed that there was some kind of interaction between the metal and the surrounding gas which resulted in the ionization of the latter. The unipolarity of the currents was explicable as arising either from the difference in velocity of the ions of opposite sign or from a difference in their chemical affinity for the hot metal or, possibly, from a combination of these causes. On such a view the detailed investigation of the mechanism of the electrical conductivity and the determination of the nature of the ions became of the utmost importance.

PROPERTIES OF THE GASES DRAWN AWAY FROM THE NEIGHBOURHOOD OF INCANDESCENT BODIES.

The nature of the electrical conductivity exhibited by gases drawn from the neighbourhood of hot wires was investigated by McClelland.[1] In many respects the phenomena were found to be similar to those exhibited by gases which had been exposed to the action of Roentgen or Becquerel rays. Thus, in examining the relation between the current and the applied

[1] "Phil. Mag.," Vol. XLVI, p. 29 (1899); "Camb. Phil. Proc.," Vol. X, p. 241 (1899); Vol. II, p. 296 (1902).

electromotive force, between suitable electrodes immersed in such gases, McClelland found that for sufficiently small differences of potential the currents were proportional to the applied potential differences. As the potential differences increased, the rate of increase of the current fell off until finally a stage was reached when the current acquired a constant maximum value independent of further increase in the potential difference. In these experiments the gases were allowed to stream at a constant rate through the testing vessel, and the maximum or *saturation* current was interpreted as indicating that all the ions present in the gas at entering the vessel were drawn to the electrodes by the electric field. This inference was established by allowing the gas to pass into a second testing vessel, when its conductivity was found to have disappeared. In these respects the gases resembled those which had been exposed to Roentgen rays and other ionizing agents. There were, however, important differences. For example, the properties of the gas depended to a very large extent on the temperature of the hot wire. With the wire at a dull red heat the gas drawn away would discharge a negatively charged conductor but not one which was positively charged. At sufficiently high temperatures charges of either sign were discharged with about equal facility. It thus appears that at low temperatures the ions drawn away from the hot metal are all positive, whereas at higher temperatures ions of both signs are present in amounts which, if not equal, are at any rate comparable with one another. These observations are at once seen to be in agreement with those recorded by Elster and Geitel. McClelland observed the excess of positive ionization at low temperatures with wires of platinum, iron, German silver, and brass, and with carbon dioxide as well as air.

McClelland also measured the mobility of the ions, i.e. their velocity of drift under a unit electric field. The gas was allowed to flow at a known rate down the annular region between two coaxial circular cylinders maintained at a given difference of potential. The fraction of the total number of ions present collected by a known length of one of the cylinders was measured. From a knowledge of this fraction, which obviously increases with the mobility of the ions, the value of the mobility can be deduced. It was found to be about 20 per cent greater for the negative than for the positive ions. The absolute values were comparable with ·04 cm. per sec. per volt/cm., and were thus much smaller than those for the ions generated by Roentgen rays (about 1·5 cm. sec.$^{-1}$ per volt cm.$^{-1}$). Moreover, they were not constant but diminished as the distance travelled by the gas from the hot body increased; that is to say, the mobilities diminished with lapse of time and as the gas became cooler. The mobilities were also found to be diminished when the temperature of the wire was increased. The last effect is usually attributed to the loading up of the ions by the particles sputtered from the hot metal, as sputtering is known to increase rapidly with rising temperature. Cooling the gas will tend to facilitate the condensation of vapours on the ions, if any vapours are present, and lapse of time will diminish the average mobility of the ions owing to recombination, since the slower ions also recombine more slowly.

These experiments showed that the currents through gases drawn away from the neighbourhood of hot bodies were carried by ions. They did not, however, throw much light on the processes by which the ions originated in the first instance, nor, since the properties of the ions under examination were clearly changing as they were carried away from the hot body, could the nature of the ions first formed easily be inferred from those of the ions under investigation. These problems were solved by experiments of a different character.

THE SPECIFIC CHARGE (ϵ/m) OF THE IONS.

The nature of the negative ions emitted by hot bodies in a gas at a low pressure was discovered by J. J. Thomson,[1] who measured the ratio ϵ/m of their electric charge ϵ to their mass m. Thomson's experiments were made with carbon filaments and the method employed was as follows: A straight filament arranged to be heated by an electric current was ·mounted

---

[1] "Phil. Mag.," Vol. XLVIII, p. 547 (1899).

parallel to and immediately in front of a metal plate A with which one end of the filament was electrically connected. A second insulated plate B was mounted parallel to A and was connected to the insulated quadrants of an electrometer. The filament was thus in the space between the two plates, which were maintained at a difference of potential V = X*d*, where X is the electric intensity and *d* the distance between the plates. The plates and filament were enclosed in a glass tube which was exhausted until the pressure of the enclosed gas was so low that the mean free path of the gas molecules was greater than the distance between the plates. If the mean free path is large compared with the distance between the plates, the influence of the gas molecules on the motion of the ions can be disregarded. The tube was placed between coils carrying an electric current, so that the plates lay in a uniform magnetic field H whose direction was parallel to that of the length of the filament. The ions starting from the filament were thus subjected to the action of a uniform electric field perpendicular to the plates, and of a uniform magnetic field parallel to the length of the filament. If the plate A lies in the plane *x* = o and the axis of *z* is taken to be parallel to the magnetic intensity H, then Thomson[1] showed that the *x* and *y* co-ordinates at time *t* of an electrified particle, starting with zero velocity from the plane *x* = o at the instant *t* = o, would be given by

$$x = \frac{m}{\epsilon}\frac{X}{H^2}\left\{ 1 - \cos\left(\frac{\epsilon}{m}Ht\right)\right\} \qquad . \qquad . \quad (1)$$

$$y = \frac{m}{\epsilon}\frac{X}{H^2}\left\{\frac{\epsilon}{m}Ht - \sin\left(\frac{\epsilon}{m}Ht\right)\right\} . \qquad . \quad (2)$$

where *m* is the mass and *ε* the charge of one of the particles. By eliminating *t* the equation to the path can be obtained. It is found to be a cycloid in the plane perpendicular to the magnetic force. The greatest distance *d* which the particles are able to travel from the plane *x* = o is determined by the equation

$$d = 2\frac{m}{\epsilon}\frac{X}{H^2} \qquad . \qquad . \qquad . \quad (3)$$

[1] Cf. J. J. Thomson, "Conduction of Electricity through Gases," p. 112, 2nd edition.

Under these conditions, i.e. if the wire is taken to be coincident with the front of the plate A, the current received by the plate B will depend on the value of X/H^2. If X/H^2 is less than $\epsilon d/2m$, none of the ions emitted by the filament will reach the plate B, whereas if X/H^2 exceeds $\epsilon d/2m$ all of them will arrive at B. There is thus a critical value of X/H^2 for which the current from A to B jumps from zero to the maximum value. If $(X/H^2)c$ denotes this critical value evidently

$$\frac{\epsilon}{m} = \frac{2}{d}\left(\frac{X}{H^2}\right)_c \qquad . \qquad . \qquad . \qquad . \qquad (4)$$

In actual practice the current does not jump with the suddenness required by this theory. With very small values of X/H^2 the current is practically zero. In fact, recent experiments by Owen and Halsall [1] and by the writer [2] show that with a number of metals and under the best conditions the current at this stage is well under one-thousandth part of the maximum value. This state of affairs persists until at a certain stage the value of the current begins to rise with increasing X/H^2. The rate of increase of the current is small at first, rapidly becomes greater, and then falls off again; so that ultimately the current exhibits a slow asymptotic approach to the final maximum value appropriate to large values of X/H^2. This divergence between theory and experiment is probably to be attributed to the fact that the ions do not set out from the hot body with zero velocity. We shall see later that at the moment of liberation the different ions set out with velocities which extend over a wide range of values.

Although this lack of sharpness rather restricts the accuracy of this method of measuring ϵ/m, the values given by it were quite exact enough to settle the nature of the negative ions. The value given by Thomson's experiments was $\epsilon/m = 8\cdot7 \times 10^6$ in E.M. units. This number agreed quite well with the values which had been obtained shortly

---

[1] "Phil. Mag.," Vol. XXV, p. 735 (1913).
[2] *Ibid.*, Vol. XXVI, p. 458 (1913).

before by Thomson and by Wiechert for the cathode rays, by Lenard for the Lenard rays, and by Thomson for the negative ions liberated from metals by the action of ultra-violet light. Before these experiments were made, the greatest value of ϵ/m with which we were familiar was that for hydrogen, the lightest chemical atom, in electrolysis. The value for hydrogen is 9.649×10^3 in E.M. units. The value found for the negative ions coming from the carbon filament was thus about 900 times as large. The importance of these experiments can hardly be over-estimated. Taken in conjunction with other experiments which served to establish the view that the charge ϵ carried by these ions was the same as that carried by a monovalent atom in electrolysis, they showed that the negative ions now under consideration were particles of much smaller mass than the chemical atoms. In other words, they proved that the carriers of negative electricity from hot bodies were the negative electrons which are now believed to form an important part of the structure of all chemical atoms.

Later experiments have confirmed these conclusions and extended the list of substances investigated. Owen[1] using a method similar to Thomson's found the value $\epsilon/m = 5.65 \times 10^6$ for the negative ions coming from a Nernst filament. Wehnelt[2] found that for the negative ions emitted by a speck of lime on a hot platinum cathode the value of ϵ/m was 1.4×10^7. His method was different from Thomson's. He showed that when a speck of lime was placed on a hot platinum cathode it formed the source of an intense beam of negative ions. The path of this beam was made visible by the luminosity it caused in the surrounding gas. The experiment was arranged so that practically all the fall of potential in the tube occurred close to the cathode, the rest of the track of the ions being almost free from the influence of the electric field. A uniform magnetic field H was then applied, so that the lines of force were parallel to the surface of the cathode and thus perpendicular to the direction of pro-

---

[1] "Phil. Mag.," vi. Vol. VIII, p. 230 (1904).
[2] "Ann. der Phys.," Vol. XIV, p. 425 (1904).

jection of the ions. Under these conditions the path of the ions is a circle of radius

$$ r = \frac{m}{\epsilon} \frac{v}{H} \quad . \quad . \quad . \quad . \quad (5) $$

in a plane perpendicular to the magnetic intensity. v the velocity of projection of the ions is given by the equation $\frac{1}{2}mv^2 = V\epsilon$, where V is the applied potential difference; so that

$$ \epsilon/m = 2V/r^2H^2 \quad . \quad . \quad . \quad (6) $$

The radius r of the path of the ions was measured by photographing the luminous track. The writer,[1] using a method which will be described later,[2] found the following values of ϵ/m for the negative ions emitted by hot bodies: for platinum $1\cdot45 \times 10^7$ and for carbon $1\cdot49 \times 10^7$. It is probable that the differences between the values of ϵ/m found by all the foregoing observers are due to errors of experiment and that all the values are too low.

More recently a very accurate investigation has been published by Bestelmeyer,[3] who used an improved form of Wehnelt's method. He found $\epsilon/m = 1\cdot766 \times 10^7$ E.M. units. This result is to be regarded as of a far higher order of accuracy than any of the preceding ones. It is unlikely to be in error by as much as $0\cdot5$ per cent. An entirely different method which preliminary experiments indicate to be capable of high accuracy has recently been developed by Langmuir and Dushman.[4]

The value of ϵ/m for the positive ions emitted by hot bodies also was first measured by Thomson.[5] The results of the researches in this direction will be considered at length later.[6] At present we shall content ourselves with the general statement that for the positive ions the values of ϵ/m have

[1] "Phil. Mag.," vi. Vol. XVI, p. 740 (1908).   [2] P. 196, chap. VI.

[3] "Ann. der Physik," iv. Vol. XXXV, p. 909 (1911).

[4] "Phys. Rev.," ii. Vol. III, p. 65 (1914); Vol. IV, p. 121 (1914).

[5] "Conduction of Electricity through Gases," p. 217, 2nd edition (Cambridge, 1906).

[6] Chap. VI. p. 194; chap. VIII. p. 261.

always been found to be as small as those occurring in electrolysis. This shows that the positive ions liberated by hot bodies are invariably structures of atomic or molecular dimensions.

General Experimental Methods.

The methods used in investigating the dependence of thermionic currents on various physical conditions, such as the temperature of the hot body, and the pressure and nature of the surrounding gaseous atmosphere, naturally depend to a considerable extent on the properties of the substance under examination. For those substances which are available in the form of wires or filaments, and which conduct electricity, as well as for numerous other substances which, owing to the magnitude of the effects to which they give rise, can be tested in the presence of a hot metal on whose surface they are deposited, an electrical method of heating is most convenient. Numerous experiments made on different substances, and by various investigators, show that there is no considerable difference in the observed effects which arise from the employment of an electric current as the heating agent, as compared with those which arise when other methods of heating are used; provided the same temperature is attained, and the other physical conditions are identical. Perhaps the most convincing evidence in this connexion is furnished by some experiments made with lime-covered cathodes by Fredenhagen,[1] who, after setting out to prove the contrary proposition, finally concluded that the method of heating made no difference. No doubt the electric and magnetic fields due to the heating current do influence the motion of the ions to some extent, but the effects thereby arising are usually not of serious importance unless very large currents are employed.[2]

The essential features of the type of apparatus most generally serviceable are exhibited in Fig. 1. The filament A to be tested is welded to stouter leads B and C. These in turn are

---

[1] "Phys. Zeits.," Jahrg. 13, p. 539 (1912); "Leipzige Ber.," Vol. LXV, p. 55 (1913).

[2] See, however, p. 61.

welded or hard soldered to platinum wires sealed into the glass bulb D. A lies on the axis of a cylindrical electrode E of metal foil, or, preferably, gauze supported by the sealed-in lead F. The tube H enables the bulb to be exhausted and sealed off or connected to the apparatus for supplying various gases, measuring the pressure, etc. The precise construction of such a bulb depends on the nature of the substance A experimented with. If A is a platinum wire then all the metal parts inside the bulb are best made entirely of platinum. The whole apparatus can then be thoroughly cleaned with boiling nitric acid and distilled water. Tungsten filaments may be clamped, or electrically welded in an atmosphere of hydrogen, to the stout leads which may be of iron or copper. Carbon filaments have to be joined with paste as in constructing incandescent lamps. Most other materials are to be welded to the supports if possible, otherwise hard soldering may be employed. In experiments of this character it is often of the utmost importance not merely to secure the chemical purity of the materials used, but to make sure that not even the smallest traces of gases are liberated in the bulb during the course of the experiments. The best way of accomplishing this is to heat the

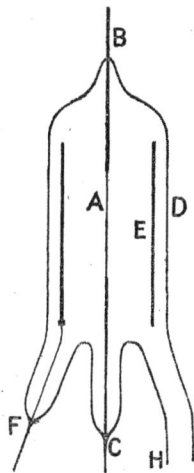

FIG. 1.

tube D to a high temperature whilst it is exhausted by a Gaede pump, assisted by a liquid air and charcoal condenser. Meanwhile the wire A is glowed out electrically, and, in order to drive every trace of gas out of the cylindrical electrode E, it is desirable that this should be heavily bombarded by cathode rays, obtained by applying a high negative potential to A. To maintain the tube D at a high temperature without its collapsing under the external pressure during the exhaustion, it should be heated in a vacuum furnace. A suitable form of furnace[1] may be constructed with a heavy water-jacketed brass

---

[1] Cf. I. Langmuir, U.S. Patent 994,010.

base provided with holes for the tube H and the leads B, C, and F. The holes can be made airtight with glass and sealing-wax, and an additional hole for the insertion of a platinum thermometer or thermocouple is desirable. On the base rests a large cylindrical brass bell jar, the line of contact being made airtight with a rubber gasket. The brass cylinder is balanced by weights attached to ropes passing over pulleys so that it can easily be moved up and down. The furnace itself is inside the brass cylinder, and rigidly attached to it. It consists, starting from the inside, of a verticle cylinder of some non-oxidizable metal such as monel metal or nickel; this is insulated by a layer of mica, over which is a winding of nichrome strip having a suitable resistance and current-carrying capacity. Between the nichrome winding and the outer brass cylinder is a thick packing of fireclay and asbestos. The leads to the nichrome strip and the exhaust can be let in through the cover of the brass cylinder. This, as well as the brass base, should be water cooled. With such an arrangement, with the furnace exhausted to a pressure of about 1 cm., the experimental bulbs can be exhausted for several days at a temperature of about 570° C. without collapsing. A vacuum furnace of this type in actual operation is shown in Fig. 2.

A method commonly employed, especially in the manu-facture of commercial electron devices, is to heat the tube to about 400° C. in an ordinary furnace and the electrodes to a much higher temperature by intense electron bombardment. The gases evolved are not allowed to accumulate but are removed by the use of very rapidly acting pumps such as mercury vapour pumps. The bombardment is stopped from time to time to allow the evolved gases to be carried away. The disadvantage of this method is that some of the material of the cathode filament is apt to be consumed by the gases evolved during the exhaustion process.

Many experiments can be made without taking these elaborate precautions, but we shall see later[1] that if we wish to be sure of obtaining the effects which are characteristic

---

[1] See chaps. III. and IV.

of the pure metals in the absence of a gaseous atmosphere we cannot afford to dispense with the manipulation just described.

Until the last few years the McLeod gauge has been exclusively relied on for the measurement of the small gas pressures dealt with. The McLeod gauge is satisfactory down to pressures in the neighbourhood of 10^{-7} mm. if the surfaces of the glass and mercury are clean and well dried by phosphorus pentoxide and if a liquid air trap is maintained continuously between the gauge and the rest of the apparatus to prevent access of mercury vapour from the gauge to the latter. This is sometimes a tiresome limitation and in any event the McLeod gauge is slow in its action. For these and other reasons there is at present a tendency to replace it by two different types of manometer, particularly in work at the lowest attainable pressures, which are of the order 10^{-9} to 10^{-10} mm. The first of these is the radiometer gauge invented by Knudsen.[1] In this instrument the deflexion of a delicately-suspended vane is measured due to bombardment by molecules of the gas recoiling from fixed parallel platinum strips which can be heated to various temperatures. The advantages of this instrument lie in its wide range of sensibility, quickness of action, and the fact that the pressures can be calculated from the dimensions of the apparatus and the knowledge of the nature of the gas present without specially calibrating the instrument. The second device is the ionization gauge. Although considerably older as a device for detecting small quantities of gases in vacuum tubes, its specific development as a pressure gauge is due to Buckley.[2] This device consists of an electron emitting cathode which we shall take to be at zero potential, an anode at a positive potential considerably in excess of the ionizing potential of the gas (for example, at a potential of 200 volts), and a third electrode at a small negative potential (for example, 10 volts).

[1] "Ann. der Physik," Vol. XXXII, p. 809 (1910); Vol. XLIV, p. 525 (1914); cf. also Woodrow, " Phys. Rev.," Vol. VI, p. 491 (1914), and Sherwood, " Phys. Rev.," Vol. XI, p. 241 (1918).

[1] " Prac. Nat. Acad. Sci.," Vol. LXVIII, p. 3 (1917); cf. also Dushman and Found, " Phys. Rev.," Vol. XV, p. 133 (1920).

If there is any gas present in the device it is ionized by the electron stream in the main discharge from anode to cathode and the positive ions produced flow towards the third electrode. The current in a line joining the third electrode to the cathode is jointly proportional to the gas pressure and to the current in a line joining cathode and anode. Thus the ratio of the two currents is equal to a constant multiplied by the gas pressure for a given structure and voltages. Wehnelt cathodes are in general more suitable than tungsten filaments as the latter attack and "clean up" all gases except those of the argon group. The advantages of this device are its wide range of sensibility, convenience, rapidity of action, and robustness. It requires a preliminary calibration at a known pressure. Strictly speaking, it measures the number of molecules per c.c. and not the pressure. This is an advantage for many purposes.

Almost all the phenomena under consideration are very sensitive to small changes in temperature; so that even when it is not necessary to know the actual temperature of the filament A it is essential that it should not vary. A very sensitive temperature control is provided by a method which involves the measurement of the resistance of the filament. For this purpose, in carrying out the experiments, the filament is made to form one arm of a Wheatstone's bridge which is actuated by the battery supplying the heating current. The arrangement of apparatus for measuring the thermionic current which flows from the filament A to the cylinder E is shown diagrammatically in Fig. 3. K, L, and M are the three other resistances which form the arms of the Wheatstone's bridge, the bridge galvanometer G being provided with the key N. The main heating current is supplied by the battery P and regulated by the system of rheostats Q, R, S. In these experiments a very fine adjustment of the current is necessary. This is supplied by placing two of the rheostats R and S in parallel. Then if, for example, the total resistance of R is very much larger than that of S a displacement of the slide wire of R will make very little difference to the total resistance of the combination R, S. In this way any degree of fineness of regulation is obtainable. Since A is to be heated to a high temperature it is necessary that a large current should flow

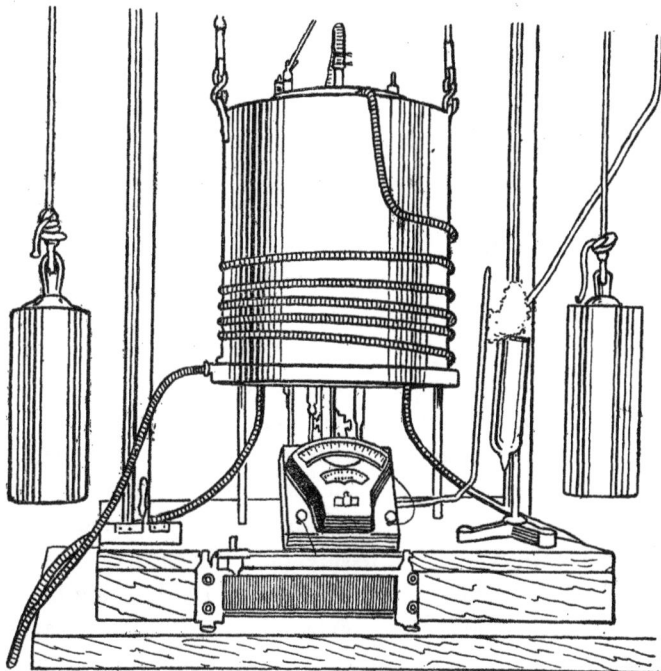

FIG. 2.

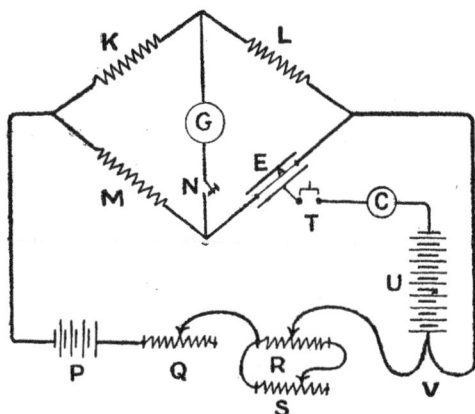

FIG. 3.

through it. Thus M must be a resistance comparable in

magnitude with A, and capable of carrying a large current without heating. If then K and L are both large compared with M and A, practically all the current will flow down the arms M, A, and the arms K, L will not be in danger of over-heating, even when the bridge is adjusted. Although there is a great disparity in the resistances of adjacent arms of the bridge this arrangement is a very sensitive tempera-ture indicator on account of the very large currents which flow down the arms A, M. In making observations at a constant temperature the bridge is adjusted initially and the galvanometer spot is kept on the zero subsequently by altering the controlling resistances Q, R, S. It is desirable to provide a shunt for the bridge galvanometer G as the currents through it may be quite large before the final adjustments are made.

In order to measure the thermionic current the cylinder E is connected to a point V in the heating circuit through the battery U, the switch T, and the current-measuring instrument C. The battery U is required, in general, to drive the ions across the gap AE. The nature of C depends on the mag-nitude of the currents to be measured. If these are large an ordinary galvanometer or even a millammeter may be used, but for the small currents obtainable at low temperatures an electrometer has to be employed. The writer has found[1] a very convenient and universal arrangement to consist of a quadrant electrometer set up so that various capacities or resistances can be thrown in across the quadrants. For the smaller currents the time rate of deflexions are then measured either with or without additional capacity across the quadrants. For the larger currents the steady deflexions across the re-sistance are observed.

For measuring the temperatures of the filaments various methods have been employed. For those materials, such as platinum and tungsten, whose resistance as a function of temperature is known with sufficient accuracy it is most con-venient to deduce the temperatures directly from the measured values of the resistance. For exact work it is necessary to

---

[1] " Phil. Mag.," Vol. XXII, p. 675 (1911).

take account of the fact that the temperature falls off towards
the ends of the filament, and is uniform only in the middle.
This difficulty can be overcome if the cylinder E is divided by
horizontal sections into three separate parts, the upper and
lower cylinders then functioning as guard rings.   It is also
important that the resistance-temperature calibration should
be made under the conditions of temperature distribution
obtaining in the experiments.   This can be secured by placing
minute fragments of salts of known melting-point on the
central portion of the wire or filament after it has been removed
from the experimental tube.   The wire is then heated electrically in a suitable atmosphere and the resistance at which the
salts melt determined.   The observation of the melting of the
salts is made with a low-power microscope.   It is desirable
that some of the salts chosen should have their melting-points
in the temperature region under investigation.   The pieces of
salt should be very minute, otherwise their temperature will
not be the same as that of the filament on which they are
placed.   In some cases small bits of fine metal wire or foil can
be used instead of salts.   A list of fixed temperatures which
are often useful in this kind of work is given in the following
table.   The melting-point of tungsten is the result of an accurate determination by Langmuir[1] and that of molybdenum
is due to Mendenhall and Forsythe.[2]   The remaining temperatures above the melting-point of iron ($1503°$ C.) are taken
from the "Recueil de Constantes Physiques," published by
the French Physical Society in 1913, the others are taken
from a list of reliable fixed points supplied by Dr. J. A.
Harker to the Cambridge Scientific Instrument Company :—

Temperature of—	Degrees Centigrade.
Liquid hydrogen	− 253
,, oxygen	− 182
Melting ice	0
Boiling-point of water at 760 mms. pressure . .	100
,, of naphthalene at 760 mms. pressure . .	220
Melting-point of tin	232
,, of lead	327
,, of zinc	419

[1] "Phys. Rev.," Vol. VI, p. 138 (1915).
[2] "Astrophys. Jour.," Vol. XXXVIII, p. 196 (1913).

2 *

Temperature of—	Degrees Centigrade.
Boiling-point of sulphur at 760 mms. pressure . .	445
Melting-point of aluminium	657
,, of sodium chloride	800
,, of silver (in air)	955
,, of silver (in reducing atmosphere) .	962
,, of gold	1064
,, of K_2SO_4	1070
,, of nickel	1427
,, of iron	1503
,, of palladium	1542
,, of platinum	1755
,, of zirconium	2350
,, of iridium	2360
,, of molybdenum	2535
,, of tantalum	2798
,, of tungsten	3267 ± 30

The resistance method of deducing the temperature has the advantage that it does not introduce any complications into the experimental bulbs. On the other hand, it often involves a separate research into the resistance-temperature relations of each substance investigated, and moreover, the resistance of some substances is not a sufficiently definite or sensitive function of temperature. The method of most general applicability is the thermocouple, but this is difficult to employ in the case of fine filaments on account of the local reduction of temperature caused by its presence. In any event the thermocouple should be made of very fine wire, and a calibration under conditions as near as possible to the experimental should be carried out, as with the resistance method. The couple should be welded to the centre of the hot wire if this is possible, or it may be cemented with platinum chloride solution, afterwards converted into platinum by heating.[1] For temperatures up to about 1500° C. couples of platinum and the alloy 90 per cent platinum + 10 per cent rhodium are satisfactory. For still higher temperatures it is probable that tungsten-molybdenum couples could be used.

Where these electrical devices are inapplicable, optical methods may be employed. The easiest of these methods is to compare the intrinsic brightness of the filament with

[1] Deininger, " Ann. der Phys.," iv. Vol. XXV, p. 292 (1908).

that of a second filament, used as an intermediate standard, whose brightness is regulated by controlling the power supplied to it. The intermediate standard is finally calibrated against a surface of the same material heated in a furnace to known temperatures. Up to the present optical methods have not been much used in this kind of work.

The writer[1] has pointed out that some of the thermionic properties of bodies are well adapted for development into thermometric methods at high temperatures, but various causes have retarded their successful exploitation up to the present.

As an illustration of the application of the thermocouple method reference may be made to an apparatus used by Deininger.[2] This apparatus represents several variations from Fig. 1, which are of advantage for special purposes. The upper lead B of Fig. 1 is bent round inside the bulb so as to pass downwards outside the cylinder E and come out of the bottom of the bulb alongside C. The two leads of the thermocouple are also brought down to the bottom of the bulb. All four leads are mounted in a stopper which is fitted to the bulb by a mercury-sealed ground glass joint. The cylinder E is provided with a vertical slit through which the wire A can pass. Thus the whole system of filament and thermocouple can be withdrawn from the apparatus. In order readily to interchange the filaments the welded joints between A and B and A and C respectively are replaced by small screw clamps. This type of arrangement has obvious advantages where it is desired rapidly to change the material under investigation. On the other hand, it has not been found feasible, up to the present time, completely to eliminate traces of gas from apparatus containing greased ground glass joints; if sealing-wax and high speed pumps are used, however, this difficulty can be overcome.

A convenient arrangement for making rapid qualitative tests of the sign of the ions emitted by hot wires has recently

[1] "Phys. Rev.," i. Vol. XXVII, p. 183 (1908).
[2] "Ann. der Phys.," iv. Vol. XXV, p. 288 (1908).

been described by Hopwood.[1] This consists in bring-
ing an electrically-charged rod near a long loop of the
electrically-heated wire which is connected to earth. The
loop will only be deflected provided it does not emit ions
of the sign opposite to the charge on the rod.

RELATION BETWEEN THE CURRENTS AND ELECTRO-MOTIVE FORCE AND GAS PRESSURE.

The first experimental investigation of this question was
made by McClelland [2] with an arrangement similar to Fig. 1
in its main features. In all these experiments the temperature of the filament is kept constant and the general character
of the results to be described is independent of the nature
of the material used, provided that the filaments have been
heated for a considerable length of time (see pp. 66 and 199).
At pressures comparable with atmospheric the relation between
the current and the difference of potential maintained between the filament and the cylinder is similar to the left-hand
half of the curve shown in Fig. 4, although this figure actually refers to another case. With low voltages the current
is proportional to the applied potential difference, but as the
potential difference increases, the rate of increase of the
current gradually falls off until finally saturation is attained.
There is thus a definite limit to the number of ions liberated
by the glowing filament in unit time. In air at low temperatures this description applies only when the filament is positively charged; there is no appreciable current when the wire
is charged negatively. At higher temperatures similar results
are obtained whether the filament is charged positively or negatively, the ratio between the saturation current with the wire
negative and that with the wire positive increasing continuously
with rising temperature. In these respects the observations
agree with the earlier experiments described on page 3.

With pressures of the order of 1 millimetre of mercury the
current-E.M.F. curves were found to be entirely different.

---

[1] "Phil. Mag.," Vol. XXIX, p. 362 (1915).
[2] "Camb. Phil. Proc.," Vol. XVI, p. 296 (1901).

With the filament negatively charged there was no indication of saturation. The current in general increased more rapidly than the first power of the potential difference. The effects observed with a *positively* charged wire at these pressures are exhibited in Fig. 4, which actually refers to this case. At intermediate voltages there is evidence of saturation, but this stage is succeeded by a stage in which the current again increases with rising potential difference.

McClelland showed that these phenomena could be explained in a general way on the hypothesis that the ions liberated at or near the surface of the filament were able, under the accelerating influence of the electric field, to produce new ions by impact with the neutral molecules with which

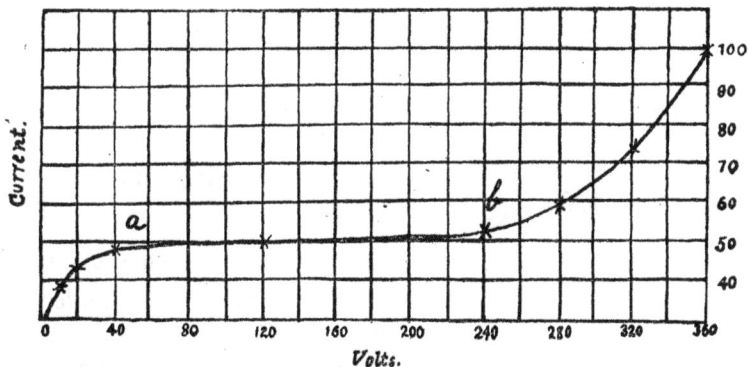

FIG. 4.

they collided. In the case of the positive ions this increase in the current due to ionization by impact did not begin to make itself felt, in the example shown in Fig. 4, until there was a difference of potential of over 200 volts between the electrodes. The existence of saturation with lower potentials showed that all the ions initially liberated were being collected by the cylinder, and, as the current was independent of the electromotive force in this region, there was no additional current depending on the energy of the impacts. The absence of saturation with the currents from the negatively charged wire makes it necessary to suppose that ionization by collision sets in before the stage at which saturation is reached. Thus the part *a, b* of the curve in Fig. 4 is missing when the wires are charged negatively. Another consequence of this interpretation is that ionization by impact is effective with a smaller fall of potential for negative than for positive ions. The hypothesis of ionization by impact had previously been put forward by Townsend and J. J. Thomson to account for somewhat similar phenomena exhibited by other sources of ionization. McClelland's experiments seem, however, to have first indicated definitely that positive ions could give rise to new ions by collision.

A more detailed examination of the relation between current, pressure, and electromotive force with negatively charged wires has been made by H. A. Wilson.[1] Some of the results obtained by Wilson at pressures ranging from 0·0036 mm. to 760 mm. are shown in Figs. 5 and 6. If the increase of current beyond the horizontal line corresponding to the lowest pressure (0·0036 mm.) is due to ionization by collision, Wilson showed that, according to Townsend's theory,

$$\log n_b/n_a = \frac{V}{E \log {}^b/_a} \left\{ e^{-(NEpa/V)\log {}^b/_a} - e^{-(NEpb/V)\log {}^b/_a} \right\} \quad (7)$$

provided $V/pa\log {}^b/_a$ is greater than 200. In this equation n_b is the number of negative ions which reach the cylinder in unit time, n_a is the number emitted by the hot wire (of circular section) in the same time, V is the applied potential difference, p the gas pressure, b the radius of the cylinder, a, that of the wire, N the number of ionizing impacts per moving ion per cm., and E the potential fall necessary to acquire the ionizing energy (ionizing potential). This formula was found to represent the experimental results satisfactorily, with N = 3·04 and E = 17·7 (volts). The values of the constants are in satisfactory agreement with those deduced by Townsend from experiments with ionized air at ordinary temperatures, when allowance is made for the difference in the number of molecules in unit volume of a gas at a definite pressure due to change in temperature. The results point to the rather important conclusion that ionization by impact depends solely

[1] "Phil. Trans., A.," Vol. CCII, p. 243 (1903).

on the nature of the molecules, and their distance apart, and
has nothing directly to do with the temperature of the gas.

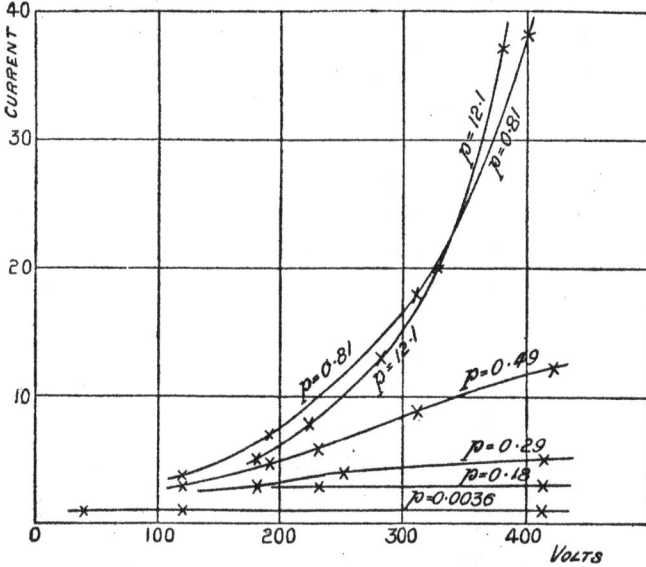

FIG. 5.

Referring to Figs. 5 and 6, we see that both at very high
and at very low pressures the current is independent of the

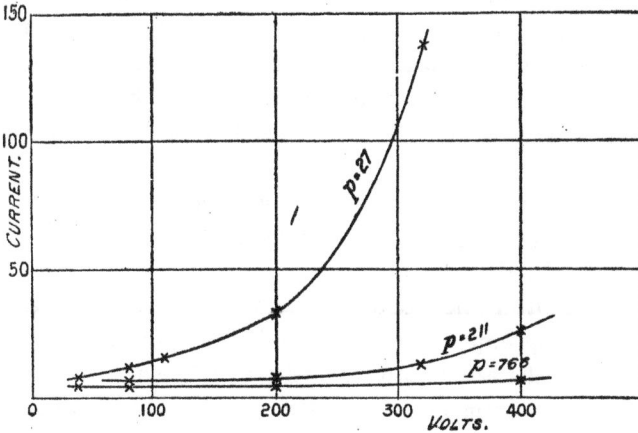

FIG. 6.

electromotive force except at the lowest voltages.   In the
former case the molecules are so crowded together that the
ions never move freely long enough to acquire the energy

necessary for impact ionization: in the latter case there are no molecules to collide with. Thus ionization by collision will occur only over an intermediate range of pressures whose extent is determined by the magnitude of the applied potential difference. In fact, if we maintain a constant potential on the filament, and gradually increase the pressure, starting from zero, the current will increase to a maximum value, and then fall off again. This experiment was made by Wilson, who showed that the pressure for the maximum current agreed with the value calculated from equation (7).

A series of observations of the relation between the currents, with the wire *positively* charged, and the electromotive force, at different pressures, was made by the writer,[1] using a platinum wire in an atmosphere of oxygen. The curves are similar to those shown in Figs. 5 and 6, except that, at a given pressure, the potential difference at which ionization by collision begins to make itself felt is much higher than when the wire is charged negatively. The increase of the current to a maximum value at intermediate pressures when the applied potential difference was kept constant was also observed when the wire was charged positively. These results could be explained by the theory of ionization by collisions on the assumption that positive ions, as well as negative, were effective, and led to an estimate of the magnitude of the ionizing energy for the positive ions from hot bodies similar to that which had been deduced by Townsend for the positive ions set free in gases by other agencies.

Recent experiments by Pawlow[2] and by E. v. Bahr and J. Franck,[3] using a more direct method, have led to further information as to the impact ionization caused by the positive ions from hot bodies. According to these authors ionization by impact sets in at ionizing potentials which are practically the same both for positive and negative ions. At these low potentials, however, the positive ions are comparatively inefficient, and their ionizing power only becomes com-

---

[1] " Phil. Trans., A.," Vol. CCVII, p. 8 (1906).

[2] " Roy. Soc. Proc., A.," Vol. XC, p. 398 (1914).

[3] " Verh. der Deutsch. Physik. Ges. Jahrg.," 1914.

parable with that possessed by the negative ions at much higher potentials. It is this last-named property which accounts for the observations recorded by McClelland and the writer. The reliability of the conclusions drawn by v. Bahr and Franck and by Pawlow has recently been called in question by Horton and Davies, who were unable to detect impact ionization in helium with 200-volt positive ions.[1]

THE ELECTRON THEORY.

We have seen that in 1899 Thomson showed that the negative ions liberated from a hot carbon filament at a low pressure were electrons. About that time a considerable amount of evidence had been accumulated which indicated that with progressively increasing temperatures and diminishing pressures, the proportion of the number of negative to the number of positive ions liberated at the surface of hot metals became increasingly greater. McClelland[2] showed further that at fairly low pressures the currents from a negatively charged platinum wire were influenced little, if at all, by changes in the nature and pressure of the surrounding gas. At the same time the electron theory of metallic conduction had made considerable advances owing to the researches of Thomson,[3] Riecke,[4] and Drude.[5] According to this theory the conductivity of metals arises from the presence in them of an atmosphere of electrons. These are supposed to be in violent motion like the molecules of a gas according to the kinetic theory of gases. The effect of an applied electric field is to superpose on the haphazard heat motion of these electrons a definite average velocity of drift in the direction of the electric potential gradient. This drifting of the electrons constitutes the electric current. The energy of the heat motion of these

[1] " Roy. Soc. Proc., A.," Vol. XCV, p. 333 (1919).

[2] " Camb. Phil. Proc.," Vol. X, p. 241 (1900).

[3] " Applications of Dynamics to Physics and Chemistry," p. 296, London (1888) ; Congrès Int. de Physique, Paris (1900); " Rapports," Vol. III, p. 138.

[4] " Ann. der Phys.," Vol. LXVI, pp. 353, 545, 1199 (1898); Vol. II, p. 835 (1900).

[5] *Ibid.*, Vol. I, p. 566 ; Vol. III, p. 369 (1900).

internal "free electrons," as they are often called, will increase
with rising temperature; and one might expect that at sufficiently high temperatures this energy would be great enough
to carry them out through the surface of the hot body.   Under
these conditions the body would be capable of discharging
negative but not positive electricity and the expected phenomena would be similar to those which appeared to characterize
the discharge of negative electricity from hot bodies, so far as
they were then known.   The probability of such a view ultimately proving correct was pointed out by Thomson [1] in 1900.
From this standpoint the escape of negative electricity from
hot bodies is closely analogous to the escape of the molecules
of a vapour from a solid or liquid during evaporation.   It is,
in fact, a kind of evaporation of electricity.   The first calculations of the thermionic currents to be expected at different
temperatures, on the view that the discharge from a negatively
charged conductor was carried by electrons shot out owing to
the vigour of their heat motions, were given by the writer,[2]
who also adduced fresh experimental evidence in support of
his conclusions.   The theory of these effects will be considered
at length in the next chapter; but for the sake of brevity and
clearness the historical order of development will no longer be
strictly followed.

[1] " Paris Rapports," Vol. III, p. 148.
[2] " Camb. Phil. Proc.," Vol. II, p. 286 (1901); " Phil. Trans., A.," Vol. CCI,
p. 497 (1903).

CHAPTER II.

THERMODYNAMICAL CONSIDERATIONS.

THE experiments recorded in the last chapter, and others to be described later, show that electrons are continually being emitted by hot solids even in a good vacuum. Consider the case of a hot solid or liquid, whose vapour pressure is negligible, contained in an exhausted vessel whose walls are insulators of electricity, the whole system being maintained at a uniform temperature T. Then there will be an accumulation of electrons in the vacuous space arising from the emission referred to. This accumulation will not go on indefinitely because some of the electrons, on account of their heat motion, will always be returning to the hot body from which they started. In consequence of these two processes a balance will ultimately be established when as many electrons return to the hot body as are emitted from it in any given interval. In this steady state there will be a definite number n per unit volume, on the average, in the vacuous enclosure, and they will exert a definite pressure p. If the enclosure is provided with a cylindrical extension in which an insulating piston can move backwards and forwards, this pressure p can be made to do work against an external force. For simplicity we may suppose that the walls of the enclosure and the cylinder and piston do not emit any appreciable number of electrons at the temperature under consideration. They are to be regarded simply as geometrical boundaries impervious to electrons.

The relation between the pressure of these electrons and the temperature of the enclosure can be found by an application of the second law of thermodynamics. The advantages of this method are that the results are independent of any suppositions about the condition of the electrons inside the hot body, and that the conclusions arrived at will possess a degree of certainty attainable in no other way, inasmuch as the second law of thermodynamics is one of the very few principles in physics to which there are no exceptions.

We know that it follows from the second law of thermo-dynamics that the entropy S of any system is a complete differential when T and p or T and v, where v is the volume of the system, are taken, respectively, as pairs of independent variables. For our present purpose a knowledge of the total entropy S of the system is not required. All we need is an expression for dS, the increment in the entropy caused by a motion of the piston. If ϕ is the change in the energy of the system which accompanies the transference of each electron from the hot body to the surrounding enclosure, then

$$dS = \frac{1}{T}\Big\{d(nv\phi) + pdv\Big\}$$

$$= \frac{1}{T}\Big\{\Big(p + n\phi + v\frac{\partial(n\phi)}{\partial v}\Big)dv + v\frac{\partial(n\phi)}{\partial T}dT\Big\} \quad . \quad . \quad (1)$$

Thus $\Big(\frac{\partial S}{\partial v}\Big)_T = \Big(p + n\phi + v\frac{\partial(n\phi)}{\partial v}\Big)\Big/T \quad . \quad . \quad . \quad (2)$

$$\Big(\frac{\partial S}{\partial T}\Big)_v = \frac{v}{T}\frac{\partial(n\phi)}{\partial T} \quad . \quad . \quad . \quad . \quad . \quad (3)$$

By equating the values of $\frac{\partial^2 S}{\partial v \partial T}$ given by (2) and (3), we find

$$T\frac{\partial p}{\partial T} = p + n\phi \quad . \quad . \quad . \quad (4)$$

since $\frac{\partial(n\phi)}{\partial v} = o$ unless the piston is quite close to the emitting surface. Now the pressure p exerted by the electrons on the piston will be the same as that exerted at the same temperature by a perfect gas having the same number of molecules in unit volume. In the case that we are considering it is to be remembered that the concentration n is so small that effects arising from the mutual repulsions of the electrons are negligible. Thus

$$p = nk\mathrm{T} \qquad . \qquad . \qquad . \qquad . \qquad (5)$$

where k is the gas constant reckoned for a single electron. By substituting the value (5) in (4) we find

$$\frac{dn}{n} = \frac{\phi}{k\mathrm{T}^2}d\mathrm{T} \qquad . \qquad . \qquad . \qquad (6)$$

or $$n = A e^{\int^{\mathrm{T}} \frac{\phi}{k\mathrm{T}^2} d\mathrm{T}} \qquad . \qquad . \qquad (7)$$

where A is independent of T.   We have thus found a relation between the number, per unit volume, of the electrons which are in equilibrium with the hot body at a point not too near its surface, and the change of energy which occurs when an electron is emitted by the hot body.

In the experiments on thermionic currents we do not measure the number n of electrons in equilibrium with a hot body but the number N emitted by unit area of its surface in unit time.   There is, however, a simple relation between these two quantities.   We shall assume as a sufficiently close approximation to the truth for our present purpose that all the electrons which return to the hot body from the surrounding space are absorbed by it.   The limitations thus introduced will be considered later.[1]   According to the principles of the kinetic theory of gases, which there is every reason to believe will apply with exactness to the behaviour of the external atmosphere of electrons, the number N' which reach unit area in unit time is

$$\mathrm{N}' = n\sqrt{\frac{k\mathrm{T}}{2\pi m}} . \qquad . \qquad . \qquad . \qquad (8)$$

where m is the mass of an electron.   But in the state of equilibrium contemplated the number of electrons emitted by the hot body in unit time is equal to the number which return to it.   Thus

$$\mathrm{N} = \mathrm{N}' = A\sqrt{\frac{k}{2\pi m}} \, \mathrm{T}^{\frac{1}{2}} e^{\int^{\mathrm{T}} \frac{\phi}{k\mathrm{T}^2} d\mathrm{T}} \qquad . \qquad . \qquad (9)$$

[1] Cf. p. 55.

If ϵ is the electronic charge, the saturation current i per unit area of the hot body is

$$i = N\epsilon \quad . \quad . \quad . \quad . \quad (10)$$

So that if we knew the relation between ϕ and T, equation (9) would completely determine the relation between the saturation current and the temperature at all temperatures if its value at a single temperature were given.

THE RELATION BETWEEN ϕ AND T.

An approximate idea of the way in which ϕ varies with T may be obtained by a rather different thermodynamic argument. We consider[1] two conductors A and A' of the same material, each of sufficiently large size. A and A' are maintained respectively at the temperatures T and T' and are connected by a thin conductor of the same substance covered with an insulating material impervious to electrons. Each conductor is surrounded by an evacuated chamber with insulating walls, and by means of a suitable arrangement of pistons and cylinders electrons can be transferred reversibly from one chamber to another in the manner described below.

In general, although the parts A and A' are connected by a conductor their surfaces will not necessarily be at the same potential on account of the difference of temperature. Such a difference of potential may arise, for example, if the contact difference of potential of metals depends upon temperature. Let us suppose that the potentials of A and A' when connected together are V and V' respectively, and that V is greater than V'. Surround A by a surface maintained at the potential V'. The effect of this will be to reduce the pressure of the electrons from the equilibrium value p characteristic of A at temperature T to the value p_0 outside the equipotential surface referred to, where

$$\log p_0 = \log p - \frac{\epsilon(V' - V)}{kT} \quad . \quad . \quad (11)$$

Equation (11) follows from the supposition that the pres-

[1] Cf. O. W. Richardson, " Phil. Mag.," Vol. XXIII, p. 602 (1912) ; " Electron Theory of Matter," p. 448. Cambridge (1914).

sure of the electrons obeys the law of a perfect gas $p = nk\mathrm{T}$. No work against the electrical forces will now be done if we remove some of the electrons which have passed through the equipotential surface from the chamber surrounding A to that surrounding A′.

Now suppose that N_o electrons are taken out of A (Fig. 7) by means of the piston and cylinder working in the walls of the surrounding chamber, at temperature T, potential V′, and pressure p_o. They are then caused to expand adiabatically to the temperature T′. The expansion is continued isothermally at T′ to the pressure $p′$, which is the equilibrium pressure of the electrons outside A′. They are then allowed

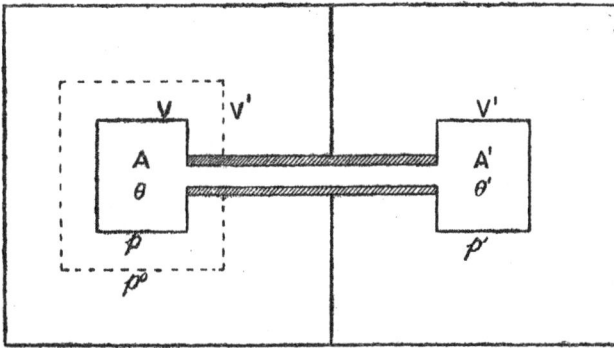

Fɪɢ 7.

to condense in A′ and finally to run down the connecting conductor to A. Since the conductor varies in temperature from point to point they will absorb heat in it to the amount $N_o\epsilon\int_{\mathrm{T'}}^{\mathrm{T}} \sigma d\mathrm{T}$, where σ is the quantity of heat liberated when unit quantity of electricity flows down unit difference of temperature under these conditions. If we apply the equation $\int\dfrac{d\mathrm{Q}}{\mathrm{T}} = o$ to this reversible cycle, we find after calculating the amount of work in each of the processes already indicated, and substituting from (11), that, if γ is the ratio of the specific heats of the electrons at constant pressure and at constant volume,

3

$$\frac{\phi}{T} - \frac{\phi'}{T'} + k \left\{ \log p - \log p' - \frac{\gamma}{\gamma - 1} (\log T - \log T') \right\}$$

$$+ \epsilon \int_{T'}^{T} \frac{\sigma}{T} dT = 0 \qquad . \qquad . \qquad . \quad (12)$$

or $\quad \log p - \dfrac{\gamma}{\gamma - 1} \log T + \dfrac{\phi}{kT} + \dfrac{\epsilon}{k} \displaystyle\int_{0}^{T} \dfrac{\sigma}{T} dT = A \quad . \quad (13)$

where A is independent of T. Differentiating this equation by T and substituting from (5) we find

$$\frac{1}{n} \frac{\partial n}{\partial T} = \frac{1}{\gamma - 1} \frac{1}{T} + \frac{\phi}{kT^2} - \frac{1}{\partial T} \frac{\partial \phi}{\partial T} - \frac{\epsilon}{k} \frac{\sigma}{T} \quad . \quad (14)$$

But from (6)

$$\frac{1}{n} \frac{\partial n}{\partial T} = \frac{\phi}{kT^2}$$

hence $\qquad\qquad \dfrac{\partial \phi}{\partial T} = \dfrac{k}{\gamma - 1} - \epsilon\sigma \qquad . \qquad . \quad (15)$

The value of σ in these expressions will not in general be quite the same thing as the specific heat of electricity measured with voltaic currents. The reason for this is that ϕ refers to a virtual displacement of the electrons, and the conditions of motion affecting such a virtual displacement will not in general be the same as those for a steady flow.[1] However, the differences arising in this way are negligible unless the conditions affecting the motion of the electrons vary very rapidly with temperature, and, in any event, there are good reasons for believing that such differences are only capable of giving rise at the most to effects of the same order of magnitude as those arising from the specific heat of electricity. Without being able to enter into the details of the conditions, about which nothing is known definitely, affecting the motion of the electrons inside the conductor, we may conclude that σ is a quantity comparable with the measured value σ_0 of the specific heat of electricity.

Among the substances where Thomson effects have been investigated the value of σ_0 is greatest for bismuth. For this substance the value of $e\sigma_0$ is about one-tenth of $k/(\gamma - 1)$, if

[1] Cf. N. Bohr, " Phil. Mag.," Vol. XXIII, p. 984 (1912).

we take $\gamma = 5/3$. As regards the other metals σ_0 is positive for some and negative for others. It is evident that $e\sigma$ will in general be much less than $k/(\gamma - 1)$; so that the greater part of the variation of ϕ with T will be determined by the first term on the right-hand side of (15). As a first approximation then we may put $e\sigma = 0$ and

$$\frac{\partial \phi}{\partial T} = \frac{k}{\gamma - 1} = \frac{3}{2}k, \text{ or}$$

$$\phi = \phi_0 + \frac{3}{2}kT \qquad . \qquad . \qquad . \qquad (16)$$

To this degree of approximation we see from (9) that

$$i = Ne = AT^2 e^{-\phi_0/kT}, \qquad . \qquad . \qquad . \qquad (17)$$

where both A and ϕ_0 are independent of T.

The first application of the principles of thermodynamics to the formation of ions by hot bodies was made by H. A. Wilson[1] in 1901. Later developments are given in papers by Wilson[2] and the writer.[3]

THE CLASSICAL KINETIC THEORY.

According to a well-known theorem of the kinetic theory of gases, there is a simple relation between the number of molecules per unit volume at any two points of a system at a uniform temperature and the work required to displace a molecule from one point to the other. Applying this theorem to the case now under consideration, it follows that if n_1 is the number of free electrons in unit volume of the interior of the hot body, the notation being otherwise as before,

$$n = n_1 e^{-\phi/kT} \qquad . \qquad . \qquad . \qquad (18)$$

Combining this result with the relation already obtained between the number N emitted in unit time and the number n in unit volume of the space outside the hot body in the state

[1] " Phil. Trans., A.," Vol. CXCVII, p. 429 (1901).

[2] *Ibid.*, Vol. CCII, p. 258 (1903); " Phil. Mag.," Vol. XXIV, p. 196 (1912).

[3] " Jahrbuch der Radioaktivitaet," Vol. I, p. 302 (1904); " Phil. Mag.," Vol XXIII, pp. 601, 619 (1912); *ibid.*, Vol. XXIV, p. 740 (1912); *ibid.*, Vol. XXVIII p. 633 (1914).

of equilibrium, the saturation current per unit area is given by

$$i = n_1 e \sqrt{\frac{k}{2\pi m}} \, T^{\frac{1}{2}} e^{-\phi/kT} \qquad . \qquad . \qquad . \quad (19)$$

If n_1 and ϕ are independent of T this is of the form

$$i = A_1 T^{\frac{1}{2}} e^{-\phi_0/kT}, \qquad . \qquad . \qquad . \quad (20)$$

and if n_1 is proportional to $T^{3/2}$ and ϕ is independent of T,

$$i = A_2 T^2 e^{-\phi_0/kT} \qquad . \qquad . \qquad . \quad (21)$$

In these equations $A_1 A_2$ and ϕ_0 are, under the suppositions named, independent of T.   Equation (21) is of the same form as (17).   Equation (19) can readily be deduced[1] by a direct calculation of the number of electrons which escape from unit area of a metal in unit time, under the supposition that the n_1 free electrons present in unit volume of the interior have a velocity distribution in accordance with Maxwell's law, and that each has to do an amount of work ϕ before it can escape from the surface.

The various calculations which have been referred to in this section all assume that the behaviour of the electrons in metals is governed by the laws of the classical dynamics. This assumption is found to lead to difficulties in other applications of the electron theory of metallic conduction.   For example, the optical properties of metals lead us to conclude that the number of free electrons present in them is quite large, and if this large number of electrons possessed the kinetic energy which the classical dynamics endows them with, the specific heats of metals would be very much larger than those actually observed.   The general course of the specific heats of metals quite precludes the view that there is any considerable number of free electrons present if the behaviour of the electrons is governed by the laws of the classical dynamics.   These are only a few of the difficulties presented by the application of the classical dynamics in this field.   It would take us too much out of our course to discuss this question at all fully. But it appears that a way of escape from most, if not all, of these difficulties opens up if we reject the classical dynamics

[1] O. W. Richardson, " Camb. Phil. Proc.," Vol. XI, p. 286 (1901).

and substitute for it the group of hypotheses, conveniently termed the quantum theory, which has recently been so successful in connexion with the theory of radiation, the properties of bodies at very low temperatures, the photo-electric effect, and the theory of the structure of atoms.

THE QUANTUM THEORY.

The bearing of the quantum theory on the emission of electrons by hot bodies has recently been considered at some length by the writer.[1] It appears that, according to the quantum theory, equation (18) is not universally true, as it is according to the classical dynamics, but is only a limit to which a more general expression approaches when the temperature becomes sufficiently high. The quantum theory is not yet completely developed, and there is a certain amount of disagreement as to the subsidiary hypotheses to be made in connexion with it. The nature of these hypotheses will affect the expression found for the general form of which (18) is a limit. The calculations are therefore to be regarded as of a provisional character, subject to possible modification as the quantum theory is developed. In the paper referred to, a calculation of the general expression corresponding to (18) has been made on the following assumptions :—

(1) That the heat energy of a gas can be analysed into the vibrations in its elastic spectrum and that the entropy of this system of vibrations can be calculated according to the method given by Planck in developing the theory of radiation ;

(2) That the elastic spectrum is limited by the number of molecules according to the principles successfully used by Debye in calculating the specific heats of solids ;

(3) That Planck's hypothesis of zero point energy has to be taken into consideration ;

(4) That the interchange of energy between gas and radiation takes place by quanta, the corresponding frequencies being twice as great in the gas as in the radiation, in accordance with the principle that the pressure exerted by a given

---

[1] "Phil. Mag.," Vol. XXVIII, p. 633 (1914).

electro-magnetic radiation has twice the frequency of that radiation ; and

(5) That the velocity of propagation of the elastic vibrations is proportional to the square root of the energy of the corresponding vibration.

The first four assumptions have been made by various writers on the quantum theory, and, so far as the writer is able to judge, have led to results in different directions which are in agreement with experience. The fifth hypothesis appears to be required to make the energy of the molecules take the equipartition value at high temperatures, and although at first sight it appears to contradict the known properties of sound waves, it is not at all certain that the contradiction is a real one. These hypotheses have been used by W. H. Keesom[1] to calculate the equations of state of gases and the thermoelectric properties of metals at very low temperatures. The results have been found to accord with the behaviour of helium at low temperatures and with the general course of thermoelectric phenomena in the same region of temperature. Moreover, a form of electron theory of metallic conduction developed by . Wien[2] along similar lines has been successful in removing a number of difficulties which the theories based on the classical dynamics were unable to overcome.

Working from the assumptions just considered, instead of arriving at (18), which may be written

$$\phi = kT \log {n_1}\Big/{n_2} \qquad . \qquad . \qquad . \quad (22)$$

we are led to

$$(w_1 - w_2)/kT =$$

$$-\frac{15}{16}(x_1 - x_2) + \frac{9}{2}\frac{1}{x_1^3}\int_0^{x_1}\frac{y^3 dy}{e^y - 1} - \frac{9}{2}\frac{1}{x_2^3}\int_0^{x_2}\frac{y^3 dy}{e^y - 1} - 3\log\frac{1 - e^{-x_1}}{1 - e^{-x_2}} \quad (23)$$

where w_1 and w_2 are the potential energies of an electron at the points in the system indicated by the suffixes 1 and 2 respectively, and x_1 and x_2 satisfy the equations

$$f(x_1) = C_1 x_1 - \frac{1}{8} \text{ and } f(x_2) = C_2 x_2 - \frac{1}{8}; \qquad . \quad (24)$$

[1] "Comm. Phys. Lab. Leiden.," Supp. No. 30 to Nos. 133-144 (1913).
[2] "Columbia University Lectures," p. 29 (New York, 1913).

in which

$$f(x) = \frac{1}{x^4} \int_0^x \frac{y^3 dy}{e^y - 1}, \quad C_1 = \frac{2}{5} \frac{MkT}{Nh^2} \left(\frac{4\pi v_1}{9N}\right)^{\frac{2}{3}},$$

$$C_2 = \frac{2}{5} \frac{MkT}{Nh^2} \left(\frac{4\pi v_2}{9N}\right)^{\frac{2}{3}} \qquad . \qquad . \qquad . \quad (25)$$

M is the molecular weight of the gas, h is Planck's constant, N is the number of molecules in one gram molecule of a gas (Avogadro's number), v_1 and v_2 are the volumes which would be occupied by one gram molecule of the gas under the concentration which it has at the points 1 and 2 respectively. The respective numbers of molecules per c.c. at these points therefore are

$$n_1 = \frac{N}{v_1} \quad \text{and} \quad n_2 = \frac{N}{v_2} k\phi$$

It is clear that the right-hand side of (23) when considered as a function of T, n_1, and n_2 will in general be quite complicated. It simplifies very considerably, however, when the quantities x_1 and x_2 are either both very small or both very large, or when one of them is very small and the other very large. It will be seen from (24) and (25) that when C is small x is large, and vice versa, and that the value of x is completely determined by that of C. The quantities N, k, and h entering into C are universal constants; so that the value of C is determined by that of the product $MTv^{\frac{2}{3}}$. It is evidently greater the greater the molecular weight of the gas, the higher the temperature and the lower the concentration. We infer from this that the behaviour of (23) appropriate to small values of C will at a given temperature occur at much smaller concentrations for an atmosphere of electrons than for an atmosphere of an ordinary gas, on account of the smallness of the mass of an electron compared with that of an atom.

When C_1 and C_2 are both large, and hence x_1 and x_2 are both small, (23) reduces, after making use of (24) and (25), to

$$w_1 - w_2 = kT \log \frac{v_1}{v_2} = kT \log \frac{n_2}{n_1} \qquad . \qquad . \quad (26)$$

This agrees with (22), since $w_2 - w_1$ is equal to ϕ for this case. Thus (22) is seen to be a limit approached by (23) for high temperatures, large molecular weights, and small concentrations. These conditions are those in which, from the point of view of this form of the quantum theory, the behaviour of gases conforms to the requirements of the classical dynamics.

When C_1 and C_2 are both small and x_1 and x_2 both large (23) again reduces to a comparatively simple expression, which, although of importance from the standpoint of the electron theory of the behaviour of metallic conductors, has no immediate application to the question now under consideration.

According to the electron theory of the optical properties of metals, the number of free electrons present in the interior of a metal is comparable with the number of atoms, and is therefore of the order 10^{23}. This conclusion is also supported by a number of other lines of argument. Now the largest thermionic currents which have been observed in a vacuum are of the order of a few amperes per square centimetre, which corresponds to an equilibrium number n of about 10^{12} at the temperature of the experiments. As a rule, n would be very much less than this. In any event, the concentrations of the external and internal electrons are seen to be of entirely different orders of magnitude. For the internal electrons in fact, C is small and x large, or at any rate approximates closely to this condition for the metals which are good conductors; whereas for the external electrons C is large and x small. We see, therefore, that it is the third of the alternatives considered above which is of interest from the standpoint of the theory of the emission of electrons from hot conductors.

In this case (23) can be shown to reduce to

$$n_2 = aT^{3/2}g(C, x_1)e^{-(w_2 - w_1)/kT} \qquad . \qquad . \quad (27)$$

where

$$a = \frac{4\pi}{9h^3}\left(\frac{3eMk}{5N}\right)^{3/2} \qquad . \qquad . \qquad . \quad (28)$$

and

$$g(C, x_1) = (1 - e^{-x_1})^3\, e^{\frac{3}{2}x_1 - \frac{2}{3}Cx_1^2}, \qquad . \qquad . \quad (29)$$

or, using (8),

$$i = Ne = \frac{2\sqrt{2\pi}}{9}\left(\frac{3e}{5}\right)^{3/2}\frac{Mk^2}{Nh^3}eT^2g(C, x_1)e^{-(w_2 - w_1)/kT}. \quad (30)$$

A numerical computation shows that over the range $1000°$ K. to $2000°$ K., $g(C, x_1)$ can be replaced without serious error by the expression $a_1 e^{b_1/T}$, where $a_1 = \cdot 473$ and b_1 is about $\frac{1}{16}$ as large as the values of the factor $(w_2 - w_1)/k$ which would be deduced if the equation (30) were applied to the experimental results given by platinum. In interpreting the results of this computation it is assumed that, over the range referred to, $n_1 = N/v_1$ can be considered to be independent of the temperature. At the higher temperatures this assumption may not be correct, and the value of a_1 would thereby be modified. In any event, a_1 does depend on the temperature (it is sensibly equal to unity at all temperatures below $1000°$ K.), and the variation of v_1 with T is not likely to affect the general character of the conclusions to be drawn. Neglecting the variation of v_1 with T it follows that the relation between the saturation current and the temperature is of the form

$$i = A_2 T^2 e^{-b_2/T} \qquad . \qquad . \qquad . \quad (31)$$

where

$$A_2 = \frac{2\sqrt{2\pi}}{9}\left(\frac{3e}{5}\right)^{3/2} \frac{Mk^2}{Nh^2} \epsilon \times \cdot 473 \quad . \qquad . \quad (32)$$

over a range from $1000°$ K. to $2000°$ K. approximately, and

$$b_2 = (w_2 - w_1 - b_1 k)/k \qquad . \qquad . \qquad . \quad (33)$$

It will be noticed that (32) is of the same form as (17) which was given as a very close approximation by the thermodynamic theory. Since the thermodynamic theory rests on considerations involving a high order of certainty, this agreement is to be regarded as a point in favour of the quantum theory. It will also be noticed that according to (32) the constant A_2 has the same value for all substances except for the comparatively small differences in the quantity a_2 which has the numerical value $0 \cdot 473$ in the particular case considered.

CONTACT DIFFERENCE OF POTENTIAL.

There is an intimate connection between the rate of emission of electrons from different substances and their contact differences of potential. This can be shown very simply by considering the case of an insulating evacuated enclosure containing two bodies A and B of different materials maintained at the uniform temperature T. The electrons emitted by A will ultimately either return to A or reach B, and vice versa. Now suppose that both A and B are uncharged initially, and that A emits electrons at a faster rate than B. The greater rate of loss of negative electrons by A will cause A to acquire a positive potential relative to that of B. This difference of potential will not increase indefinitely because the electric field thus set up will tend to stop the transference of electrons from A to B. A steady condition will finally be established in which each of the bodies A and B receives in a given time as many electrons as it emits in that time. This condition is also characterized by the occurrence of a constant difference of potential V between any two points close to the surfaces of A and B respectively. The number of electrons in unit volume of the space will then vary from point to point, but will not change with time. A consideration of the nature and number of the variables entering into the equations governing the equilibrium of the electrons[1] shows that V is independent of the size, shape, and relative position of the bodies A and B, and depends only on their nature and the temperature T. This result holds true both on the basis of the classical dynamics and on that of the quantum theory. The difference of potential V is, therefore, the intrinsic contact potential difference of the bodies A and B at the temperature T.

We have seen in the last section that on account of the small concentration of the electrons in the vacuous space outside of the emitting bodies their equilibrium will always be governed by equation (18). Thus if, in the state of equilibrium, n_1 is the number of electrons per unit volume just outside A and n_2 the corresponding number just outside B,

$$n_1 \Big/ n_2 = e^{-\epsilon V/kT} \qquad . \qquad . \qquad . \qquad (34)$$

since $_\epsilon V$ is the work done in taking an electron from a point outside B to a point outside A. If N_1 and N_2 are the numbers of electrons emitted by unit areas of the surfaces of A and B

---

[1] Cf. O. W. Richardson, " Phil. Mag.," Vol. XXIII, p. 265 (1912).

respectively in unit time, we see from equation (8), which holds true on all the theories we have considered, that

$$\frac{N_1}{N_2}\bigg|_{=}\ \frac{n_1}{n_2}\bigg|_{=}\ e^{-\epsilon V/kT} \qquad . \qquad . \quad (35)$$

or
$$V = \frac{kT}{\epsilon} \log \frac{N_2}{N_1} \qquad . \qquad . \quad (36)$$

Thus the logarithm of the ratio of the saturation currents per unit area for any two substances should vary directly as their contact difference of potential and inversely as the absolute temperature.

The contact difference of potential is also closely related to the difference in the values of ϕ, the work necessary for an electron to escape from each of the substances under consideration. Referring to equation (9), which is based on thermodynamics and therefore is independent of assumptions about the conditions affecting the electrons inside the substances, let N_1, A_1, and ϕ_1 refer to the substance A, and N_2, A_2 and ϕ_2 to the substance B, in equilibrium at the temperature T. Then by taking logarithms of the equations corresponding to (9) for each substance and subtracting we see that

$$\int^T \frac{\phi_1 - \phi_2}{kT^2} dT = \log \frac{N_1}{N_2} - \log \frac{A_1}{A_2} = -\frac{\epsilon V}{kT} - \log \frac{A_1}{A_2} \quad . \quad (37)$$

and since A_1 and A_2 are independent of T,

$$\phi_1 - \phi_2 = \epsilon V - \epsilon T \frac{\partial V}{\partial T} . \qquad . \qquad . \quad (38)$$

A similar result may be obtained by a simple application of the principle of the conservation of energy. Consider the bodies A and B to be in contact at some portion of their surface and calculate the work done in taking an element of electric charge, for example an electron, round a closed circuit partly inside and partly outside the two bodies, and passing through the part of the surface where they are in contact. The work along the part of the path outside the bodies is ϵV, the work in crossing the outside surfaces is ϕ_2 in the case of B and $-\phi_1$ in the case of A. The only work done in the part of the path inside the bodies occurs in crossing the interface and is equal to $-\epsilon P_1$ where P is the electromotive force corresponding to the Peltier effect at the junction. Since the work in traversing such a closed reversible cycle at constant temperature must be zero it follows that

$$\phi_1 - \phi_2 = \epsilon V - \epsilon P \qquad . \qquad . \qquad . \quad (39)$$

Thus we see that the second term of the right-hand side of (38) corresponds to an electromotive force equivalent to the Peltier effect at the junction between A and B. Unless the substances are very close together in the Volta series, P is small compared with V; so that the differences of ϕ_1 and ϕ_2 will be almost equal to the contact difference of potential.

To the extent to which equations (16) and (17) are valid approximations the differences of ϕ at a given temperature are equal to the differences of ϕ_0 ; so that to the same degree of approximation kT times the difference of the indices of the exponential in equation (17), which determines the temperature variation of the emission, will also be equal to the contact potential difference. By taking logarithms of (17) and subtracting we also notice that the differences of ϕ_0 are equal to $\epsilon V - kT \log A_2/A_1$; so that to the same degree of approximation it is necessary that $\log A_2/A_1 = 0$, or that A should have the same value for all substances. In dealing with the quantum theory we saw that this result was to be expected only when dealing with good conductors like the metals. This is to be expected also in the present connection because it is unlikely that with the poorer electronic conductors such as the oxides that the thermoelectric effects can be regarded as negligible. In fact S. L. Brown [1] has recorded that a copper-copper oxide couple whose junctions are at 20° C. and 530° C. respectively exhibits a thermoelectromotive force which exceeds half a volt. This means that the term $T \dfrac{\partial V}{\partial T}$ is of the same order as V in such cases.

The reader who wishes further to pursue the relation between the effects under consideration and thermoelectric phenomena may be referred to a book by the writer on the "Electron Theory of Matter," Chapter XVIII (Cambridge

[1] "Phys. Rev.," Vol. III, p. 239 (1914).

University Press, 1914). The development there given is from the standpoint of thermodynamics and the classical dynamics, but the modifications required by the quantum theory can be seen in a general way from the discussion in this and the preceding sections.

It follows from equation (35) that the relative powers of electronic emission of different bodies at a given temperature will be determined by their contact differences of potential ; so that whether bodies show much or little difference one from another in the former respect will depend on the magnitude of the latter quantity. There is still a great difference of opinion as to the magnitude of the contact difference of potential between metals whose surfaces are free from gas and in a good vacuum. The school which attributes these differences of potential to chemical action between metals and the surrounding atmosphere holds that under the conditions referred to the contact potentials would completely disappear. If this view is correct we should expect all hot metals to give nearly equal thermionic currents per unit area at any given temperature, provided they were in a perfect vacuum and their surfaces were uncontaminated. The opposite school regards these potential differences as an intrinsic property of the metals affected and considers the changes caused by gases and other contaminating agents to be of a secondary character. From this standpoint we should expect to find potential differences between metals in a good vacuum of the same order of magnitude as those observed in a gaseous atmosphere. The advocates of these opposing views have waged an intermittent warfare for a century without coming to a definite settlement.

Until recently most investigators who have attempted to decide this question experimentally have concluded that their results favoured the chemical theory. In 1912 the writer [1] pointed out that none of these experiments were conclusive, all the observed phenomena being explicable on the intrinsic theory when due account was taken of various secondary actions which were bound to occur under the conditions of the experiments. Quite recently a considerable amount of

---

[1] " Phil. Mag.," Vol. XXIII, p. 268 (1912).

evidence favouring the intrinsic theory has accumulated. Thus Richardson and Compton [1] examined the photoelectric currents obtained when monochromatic light fell on small discs of various metals placed at the centre of a large spherical electrode. With this arrangement the saturation value of the current should be reached when there is no difference of potential between the two electrodes. This was found to be the case if the contact potentials were included among the potential differences operative. Somewhat similar experiments have been made by Page. [2] In all these experiments good vacuum conditions were attained. In addition, in Richardson and Compton's experiments with sodium, and in all Page's experiments, the metal surfaces tested were cut mechanically *in vacuo*. Still more recently the contact difference of potential has been measured directly under the best vacuum conditions with surfaces machined *in vacuo* by A. E. Hennings, [3] who finds that the potential differences are still of the order usually observed, the metals being more electropositive when freshly cut. All these experiments support the intrinsic potential theory, although there is abundant evidence that gases produce definite and complicated changes in the observed values. On the other hand, Hughes, [4] working with surfaces of metals freshly distilled *in vacuo*, found the metals to be initially most electronegative and to become more electropositive under the action of small quantities of air. Millikan and Souder, [5] also, have found that surfaces of sodium are most electronegative when freshly cut and become more electropositive on oxidation. It is clear that the experimental evidence as to the origin of contact potential differences is still conflicting, but the balance would seem to favour the view that it is due to an intrinsic property of the materials and not to surface films of foreign matter.

[1] " Phil. Mag.," Vol. XXIV, p. 575 (1912); cf. also K. T. Compton, *ibid.*, Vol. XXIII, p. 579 (1912).
[2] " Amer. Jour. Sci.," Vol. XXXVI, p. 501 (1913).
[3] " Phys. Rev.," Vol. IV, p. 228 (1914).
[4] " Phil. Mag.," Vol. XXVIII, p. 337 (1914).
[5] " Phys. Rev.," Vol. IV, p. 73 (1914).

The Distribution of the Electrons in Temperature Equilibrium Outside a Hot Metal Surface.

On p. 29 we supposed that the electrons in an enclosure in thermal equilibrium containing a piece of hot metal would be distributed with uniform density except close to the surface of the metal and the walls of the enclosure. This supposition will only be valid if the concentration of the emitted electrons is exceedingly small, a condition which is satisfied in the cases considered so far. At very high temperatures the number of electrons emitted becomes very great, and then the effects which arise from their mutual repulsions can no longer be neglected. The problem which then presents itself is not merely of theoretical importance but is of considerable interest in connexion with the electrical behaviour of celestial bodies. For instance, the aurora borealis has been attributed to streams of ions from some extra-terrestrial source, probably the sun, and it is now well established that there are intense magnetic fields at the surface of the sun which are closely re-lated to disturbances in the solar atmosphere. It is natural to look to thermionic causes for the primary origin of the ionization which gives rise to these effects and the magnitude of the electrical effects which might thereby arise are seen to be of interest from the standpoint of cosmical physics.

The general condition for equilibrium in an atmosphere of electrons at constant temperature is that the force on the electrons in any element of volume arising from the electric field should balance the force on the same element of volume arising from the pressure gradient. Expressed analytically, if n is the number of electrons per unit volume, ϵ the charge of an electron, $\rho = n\epsilon$, the volume density of the electrification, E the electric intensity, p the pressure of the electrons, k Boltzmann's constant, and T the temperature, we have the following equations :—

$$n\epsilon E = \text{grad. } p = k T \text{ grad. } n = \frac{kT}{\epsilon} \text{ grad. } \rho$$

$$= \frac{kT}{4\pi\epsilon} \text{grad. div. } E = \frac{kT}{4\pi\epsilon}(\text{rot. rot. } E + \nabla^2 E) . \quad (40)$$

Thus
$$E \text{ div. } E + \frac{kT}{\epsilon} \nabla^2 E = 0 \quad . \quad . \quad (41)$$

since rot. $E = -\dfrac{1}{c}\dfrac{\partial H}{\partial t} = 0$, H being the magnetic force. In general, then, the distribution of electric intensity will be governed by the differential equation (41). If this is solved, the solution being subject to the boundary conditions of the problem, the distribution of n may be obtained from the additional differential equation.

$$\text{grad. } \log n = \frac{\epsilon}{kT} E \quad . \quad . \quad (42)$$

which is seen to follow from (40).

As an illustration we shall consider only the one dimensional case of the equilibrium of electrons in front of an emitting plane surface infinite in extent. In this case the writer [1] has shown that if v is the volume occupied by unit mass of the electrons at any point, then

$$\frac{d^2v}{d^2x} - \tfrac{1}{2}\left(\frac{dv}{dx}\right)^2 + C = 0. \quad . \quad . \quad (43)$$

where $C = 4\pi N_0^2\epsilon^2/RT$, x is the perpendicular distance from the emitting plane, N_0 is the number of electrons in unit mass (i.e. $N_0 = 1/m$ if m is the mass of an electron), and $R/N_0 = k$. The integral of (43) subject to

$$\left.\begin{array}{r}\dfrac{d \log v}{dx} = 0 \text{ when } v = \infty \\[2mm] v = \infty \text{ when } x = \infty \\[2mm] v = v_0 = \dfrac{N_0}{n_1} e^{w/RT} \text{ when } x = 0\end{array}\right\} \quad . \quad . \quad (44)$$

and

is
$$v^{1/2} = \left(\frac{2\pi}{RT}\right)^{1/2} N \epsilon x + \left(\frac{N_0}{n_1}\right)^{1/2} e^{w/2RT} . \quad . \quad (45)$$

By comparing with equation (7), p. 31, it is seen that

$$n_1 e^{-w/RT} = Ae^{\int \frac{\phi}{kT^2}dT}$$

in the notation there used. Since

$$\frac{RT}{v}\frac{dv}{dx} + N_0\epsilon\frac{dV}{dx} = 0 \quad . \quad . \quad (46)$$

[1] O. W. Richardson, " Phil. Trans., A.," Vol. CCI, p. 503 (1903).

we have if $V = V_0$ when $x = 0$

$$V = V_0 - 2\frac{RT}{N_0 e} \log \left\{ 1 + \left(\frac{2\pi n_1 N_0}{RT}\right)^{1/2} \epsilon e^{-w/2RT} x \right\} . \quad (47)$$

The electric intensity at any point x is

$$-\frac{dV}{dx} = \frac{(8\pi RT n_1/N_0)^{1/2}}{1 + (2\pi n_1 N_0/RT)^{1/2} \epsilon e^{-w/2RT} x} e^{-w/2RT} \quad (48)$$

The charge on unit area of the emitting plane is given by $\sigma = -\frac{1}{4\pi}\left(\frac{dV}{dx}\right)_{x=0}$ and the volume density at any point x is $\rho = -\frac{1}{4\pi}\frac{d^2V}{dx^2}$. It is clear that $\int_0^\infty \rho dx = \sigma$ since $\frac{dV}{dx} = 0$ for $x = \infty$. Thus the charge on the surface is equal and opposite to the total charge in the space outside.

It does not seem likely that the effects which arise in this way can be of sufficient magnitude directly to account for any important cosmical phenomena. The potential differences which develop are comparatively small. Thus, taking the case of platinum at 1500° K., if the experimental values are substituted in (47), it appears that when $x = 10$ cms. $V - V_0$ is approximately 1·5 volts, and when $x = 1$ cm. $V - V_0$ is 1·2 volts, the greatest potential gradient being at the emitting surface. These potential differences do not vary very rapidly with temperature in spite of the enormous variation of the rate of thermionic emission. Thus with platinum at 6000° C. the potential difference at a distance of 1 cm. is about sixteen times as great as at 1300° C. There is a very much more rapid change in the density of the charge on the surface which increases by a factor of about 10^8 in this interval. At 6000° C. the surface density may be of the order of 500 e.s.u. per sq. cm., and there will be an equal and opposite total charge in the overlying space.

It is not supposed that the conditions here contemplated bear any very close resemblance to those at the surface of the sun, where there is a very dense atmosphere of highly conducting hot vapours. But it would seem that the presence of this conducting atmosphere would tend further to reduce the differences of potential which arise directly from thermionic effects and to make the actual electrical effects smaller than those now contemplated.   Let us consider the magnitude of the magnetic fields which might arise from the motion in the sun's atmosphere of the electrification which arises in this way.   The magnetic intensity H at the centre of a disc charged to surface density σ and rotating with uniform angular velocity is given by

$$H = 2\pi\sigma V/c,$$

where V is the peripheral velocity and c is the velocity of light. From Hale's observations of the solar atmosphere values of V as high as 10^7 cm. per sec. appear to be permissible; so that to account for the observed magnetic intensities in sunspots (up to 3000 gauss) values of σ comparable with 10^6 e.s.u. are required.   One cannot consider values of this magnitude incompatible with the results of the foregoing calculations, since the surface density of the source of emission at any temperature depends very much on the thermionic constants of the substance.   The difficulty is rather to account for the requisite separation, in such a highly conducting atmosphere, of the positive and negative charges; since if such separation does not occur the electrifications of opposite signs will, for all practical purposes, revolve together and the resulting magnetic fields will be negligible.

In addition to the force due to their mutual repulsion, the electrons are also acted on by a force which varies inversely as the square of the distance from the emitting surface and is due to the electric charge they induce on it.   This force is independent of the concentration of the electrons in the external atmosphere and is inappreciable except at very minute distances from the surface.

The general problem of the equilibrium of electron atmospheres in cases in which the volume density effects are not negligible has recently been dealt with at considerable length by v. Laue.[1]   He points out that equations (40)-(42) can be put in the form

$$\nabla^2\psi = e^\psi \qquad . \qquad . \qquad . \quad (48\cdot01)$$

[1] " Jahrb. der Rad. u. Elektronik," Vol. XV, pp. 205, 257 (1918).

where

$$V = \frac{1}{a}(\log [a\rho_0] - \psi), \rho = \frac{1}{a}e^{\psi}, a = \frac{\epsilon}{kT} \qquad (48\cdot02)$$

and ρ is the volume density which takes the value ρ_0 at $V = 0$. Clearly the solutions of (48·01) can involve only the co-ordinates and the boundary conditions which will determine the constants of integration. The function ψ can be so determined that at the boundaries it approaches ∞ as $-2 \log x$ where x is the distance from the boundary. This is proved in a number of special cases and shown to be prob-able in general. This would make ρ infinite at the geo-metrical boundary which is physically impossible. However, if the physical boundary is an electron emitting surface and the temperature high the value of ρ appropriate to the physical boundary will be very large and the theoretical problem will correspond to the physical case in which the boundary is an electron emitting solid very close to the geometrical boundary in the theoretical problem. The distance between the two boundaries diminishes to zero as the temperature increases. Provided the temperature is so high that the two boundaries are practically coincident it follows from (48·01) that ψ is in-dependent of T and hence from (48·02) that ρ is proportional to T. In other words, in a hollow enclosure whose boundaries are maintained at a uniform temperature exceeding a limit depending on the emission constants of the surface, the equilibrium density of the electrons at points in the interior not close to the walls is directly proportional to the absolute temperature. Furthermore a consideration of the dimensions of the equations (48·01) and (48·02) shows that in similarly shaped enclosures at sufficiently high temperatures the equi-librium density of the electrons varies, at corresponding points not too near the boundaries, inversely as the square of the linear dimensions.

Two parallel plane problems are considered in addition to the one on p. 48. In the first the electrons are pressed by an opposing electric force back towards the emitting plane. In this case

4 *

$$V = \frac{I}{a}\left\{\log\left\{\sin h^2\left(\frac{I}{2}aK[x - x_0]\right)\right\} - \log\frac{Ka^2}{2\rho_0}\right\} \quad (48 \cdot 03)$$

$$\rho = aK^2/2 \sin h^2\left(\frac{I}{2}aK[x - x_0]\right) \qquad . \qquad . \qquad . \quad (48 \cdot 04)$$

where ρ_0 is the value of ρ where $V = 0$, x_0 is the boundary near the emitting plane, and K is the limiting value of the opposing electric intensity for x large. The second case[1] is that of the space between two emitting planes close to the bounding planes at x_0 and $x_0' = x_0 - \dfrac{2\pi}{aK}$. In this case

$$V = \frac{I}{a}\left\{\log\left\{\sin^2\left(\frac{I}{2}aK[x - x_0]\right)\right\} - \log\frac{aK^2}{2\rho_0}\right\} \quad . \quad (48 \cdot 05)$$

$$\rho = aK^2/2 \sin^2\left(\frac{I}{2}aK[x - x_0]\right) . \qquad . \qquad . \quad (48 \cdot 06)$$

When $K = 0$ we have the solution already dealt with, which in this notation is

$$V = \frac{I}{a} \log\left\{\left[\frac{I}{2}a\rho_0(x - x_0)^2\right]\right\}, \frac{dV}{dx} = \frac{2}{a(x - x_0)} \quad (48 \cdot 07)$$

$$\rho = 2/a\,(x - x_0)^2 \quad . \qquad . \qquad . \qquad . \qquad . \qquad . \quad (48 \cdot 08)$$

The case of spherical symmetry leads to an intractable differential equation. By omitting one of the terms a solution can be obtained but this is only of restricted validity. In the case of slightly curved surfaces, however, v. Laue is able to show that in general

$$V = \frac{I}{a} \log\left[\frac{I}{2}a\rho_0\left(\frac{I - e^{-\left(\frac{I}{R_1} + \frac{I}{R_2}\right)x}}{\frac{I}{R_1} + \frac{I}{R_2}}\right)^2\right]$$

so that

$$\frac{\partial V}{\partial x} = \sqrt{\frac{2\rho_1}{a}} - \frac{2}{a}\left(\frac{I}{R_1} + \frac{I}{R_2}\right) . \qquad . \qquad . \quad (48 \cdot 09)$$

x is the distance from, and ρ_1 the density at, the surface and R_1 and R_2 the radii of principal curvature.

[1] This has also been dealt with by Schottby, "Jahrb. der Rad. u. Electronik," Vol. XII, p. 199 (1915).

From the solutions (48·03)-(48·08) the solutions for a large number of two dimensional problems can be obtained by the method of conformal representation. These include various cases of coaxial circular cylinders, planes inclined at various angles and approximately confocal elliptic cylinders. Of these the most important from a practical standpoint is that of a small emitting circular cylinder such as a hot wire surrounded by a larger coaxial cylinder with the field so directed as to drive the electrons back towards the wire. In this case

$$V = \frac{1}{a}\left\{\log\left(\frac{a\rho_0 r^2}{8c^2}\right) + \log\left[\left\{\left(\frac{r}{r_0}\right)^C - \left(\frac{r_0}{r}\right)^C\right\}^2\right]\right\} \quad (48\cdot10)$$

$$\rho = 8C^2/ar^2\left\{\left(\frac{r_0}{r}\right)^C - \left(\frac{r_0}{r}\right)^C\right\} \qquad . \qquad . \qquad . \quad (48\cdot11)$$

$$\frac{dV}{dr} = \frac{2}{ar}\left[1 + C\coth\left(C\log\frac{r}{r_0}\right)\right] \qquad . \qquad . \qquad . \quad (48\cdot12)$$

where ρ is the density of the electrons at $V = 0$. The radius of the emitting cylinder is a little larger than r_0. The value of the constant C can be determined from the strength of the opposing field at a great distance from the emitting cylinder. For r/r_0 large this is

$$\frac{dV}{dr} = \frac{2(1 + C)}{ar} \qquad . \qquad . \qquad . \quad (48\cdot13)$$

v. Laue also finds the interesting result that with a sufficiently great emission density (such as corresponds to a high temperature, more or less, according to the electron emission constants of the substance) the layer of electrons in equilibrium with the surface will exert no pressure on it in a direction normal to the surface. They will, however, behave like a film having a negative surface tension and will exert a pressure in all directions tangential to the surface. To illustrate this we may consider the simplest case of an infinite emitting plane and no impressed external field for which the solutions are given by (48·07) and (48·08). Let the suffix 1 denote the values of the various quantities at the emitting plane; e.g. x_1, which is a little greater than x_0, is the co-ordinate of this plane, ρ_1 is the electron density at the plane and so on.   Then the electric field set up by the electron gas will give rise to a pull on each unit area of the plate in the direction of the normal which is equal to

$$\frac{1}{2}\left(\frac{dV}{dx}\right)^2_1 \text{ and this is equal to } \frac{\rho_1}{a}$$

from (48·07) and (48·08).   On the other hand the impacts of the electrons will give rise to a pressure $p = n_1 kT = \rho_1/a$. Thus these two forces just balance one another and there is no pressure normal to the surface.   In the perpendicular direction, however, the electric tension becomes a pressure and so is added to the kinetic pressure of the electron gas. The magnitude of the equivalent surface pressure (negative surface tension) per unit length will be obtained by integrating this pressure with respect to the depth of the layer of electrons. Its magnitude will therefore be given by

$$\Sigma = \frac{1}{2}\int_{x_1}^{\infty}\left(\frac{dV}{dx}\right)^2 dx + \frac{1}{a}\int_{x_1}\rho\,dx$$

$$= \frac{2}{a}\sqrt{\frac{2\rho_1}{a}} = 2\left(\frac{kT}{\epsilon}\right)^{3/2}(2\rho_1)^{1/2} \qquad (48·14)$$

Turning to equation (48·09) we see that to the first order in

$\left(\dfrac{1}{R_1} + \dfrac{1}{R_2}\right)$ at a slightly curved surface

$$\frac{1}{2}\left(\frac{\partial V}{\partial x}\right)^2_1 = \frac{\rho_1}{a} + \frac{2}{a}\sqrt{\frac{2\rho_1}{a}}\left(\frac{1}{R_1} + \frac{1}{R_2}\right) = \frac{\rho_1}{a} + \Sigma\left(\frac{1}{R_1} + \frac{1}{R_2}\right) \quad (48·15)$$

The part ρ_1/a of this tension is balanced by the kinetic pressure of the electrons; so that we see that a slightly curved electrode is subject to a capillary pressure of amount

$$-\Sigma\left(\frac{1}{R_1} + \frac{1}{R_2}\right) \qquad . \qquad . \qquad . \quad (48·16)$$

This is an outward pull if the electron emitting surface is convex and a push into the surface if it is concave.   In each case therefore it tends to increase the area of the surface.   The quantity Σ is thus strictly analogous to the surface tension of a liquid except that it acts in the opposite sense, namely, as a pressure instead of a tension. In equations (48·14)-(48·16) the quantities a, ϵ, ρ_1, and Σ are to be treated as positive numbers.

The absence of normal pressure on the boundary when the electron densities are large restricts the application of the thermodynamic calculation on pp. 29 and 32 to relatively low temperatures. It appears, however, that when due account is taken of the work done by the negative surface tension of the electron layer during the expansion a calculation along similar lines leads to the same final formula for the electron density as a function of temperature. It appears, therefore, as the kinetic theory considerations would lead us to expect, that the formulæ connecting the saturation currents and the temperature are not restricted to any particular range of temperature.

The Reflexion of Electrons from Solids.

Most of the foregoing calculations of the number of electrons emitted by hot bodies depend upon a preliminary determination of the concentration, n, of the electrons in equilibrium with the hot body in an enclosure at constant temperature. From the value of n the number N′ of electrons which return from the surrounding atmosphere to each unit of area of the hot surface is immediately deducible. Since, for equilibrium, the number N of electrons emitted must be equal to the number absorbed, we have hitherto assumed N and N′ to be equal. This will only be correct provided all the electrons which return to the hot body are absorbed by it. Experiments made by the writer (see p. 169, Chap. V) and by v. Baeyer[1] have shown that a very considerable proportion of the slowly moving electrons emitted by hot bodies is reflected from the surfaces of metals, and Gehrts[2] has shown that the same thing holds true for the electrons liberated by photoelectric action. If the proportion of the incident electrons reflected is denoted by r the correct equilibrium condition is

$$N = N'(1 - r),$$

[1] "Ber. der Deutsch. Physik. Ges.," Jahrgang 10, p. 96 (1908).
[2] "Ann. der Physik," Vol. XXXVI, p. 995 (1911).

since $N'(1 - r)$ is the number of those which are actually absorbed. For a number of metals which have been tested the value of r has been found to be in the neighbourhood of $0\cdot5$; so that the omission of its consideration, although making an appreciable change in the calculated value of N, will not alter the order of magnitude of this quantity.

LIBERATION OF ELECTRONS BY CHEMICAL ACTION.

Since it has been found[1] that electrons are emitted when various gases react chemically with the alkali metals and their amalgams, it is worth while to examine what laws an emission caused by chemical action would be expected to follow. From this standpoint the problem is a well-known one of chemical dynamics. Indications towards the solution can be obtained by the application of thermodynamics to the products of the reaction under conditions of equilibrium in an enclosure at constant temperature. The electrons and positive ions are to be regarded as products of the reaction which exert a pressure in accordance with the laws of a perfect gas. The results will be strictly accurate for very small concentrations such as correspond to thermionic emission. Let us consider first the case of gaseous reactions.

FORMATION OF IONS IN GASEOUS CHEMICAL REACTIONS.

Consider the formation of ions in a reaction in which all the products are gaseous. If the reaction is represented by the generalized chemical equation

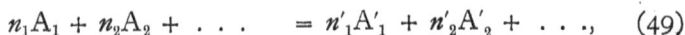

$$n_1A_1 + n_2A_2 + \ldots \quad = n'_1A'_1 + n'_2A'_2 + \ldots, \quad (49)$$

the corresponding equilibrium concentrations being denoted by the letter C with the same suffixes as the A's, it follows from thermodynamics[2] that

$$\Sigma n_s \log C_s = k \ . \qquad . \qquad . \quad (50)$$

and

$$\frac{\partial k}{\partial T} = \frac{q}{RT^2} \cdot \qquad . \qquad . \quad (51)$$

---

[1] Cf. p. 144, Chap. IV, and p. 308, Chap. IX.

[2] Cf. Van't Hoff, "Lectures on Theoretical and Physical Chemistry," Pt. I, p. 141.

where q is the latent heat of the reaction at constant volume per gram molecule and R is the gas constant per gram molecule. Integrating (51) and combining with (50), after taking out the logarithm we have

$$\Pi_s(C_s^{ns}) = e^k = e^{\int \frac{q}{RT^2} dT} \qquad . \qquad . \qquad . \qquad (52)$$

when Πs denotes the continued product of $C_1^{n_1}$, etc., the indices for the concentrations corresponding to the left-hand side of (49) being taken negative. The simplest possible reaction resulting in the liberation of ions is

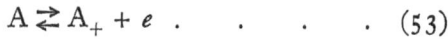

$$A \rightleftarrows A_+ + e \qquad . \qquad . \qquad . \qquad (53)$$

A being the undissociated molecule, A_+ the positive ion and e the electron. The concentrations being C, C_1, and C_2 from left to right, (52) gives

$$C_1 C_2 = C e^{\int \frac{q}{RT^2} dT} \qquad . \qquad . \qquad . \qquad (54)$$

If a is the coefficient of recombination of the positive ion and the electron and t is the time, we have from the definition of a

$$\frac{\partial C_1}{\partial t} = \frac{\partial C_2}{\partial t} = -aC_1 C_2 \qquad . \qquad . \qquad . \qquad (55)$$

Thus $a\,C_1\,C_2$ is the number of electrons which disappear in unit time. But in the steady state this must also be equal to the number of electrons liberated in unit time by the decomposition of C. Hence when the electrons are removed by an electric field as fast as they are formed, the saturation current will be

$$N\epsilon = a\epsilon C_1 C_2 = a\epsilon C e^{\int \frac{q}{RT^2} dT} \qquad . \qquad . \qquad (56)$$

This is proportional to C, as it should be. Since neither a[1] nor q vary very rapidly with T, (56) shows that the temperature variation of the currents under consideration will not be far from the form $AT^{\frac{1}{2}}e^{-b/T}$ which has been found to agree with the experiments on thermionic currents.

[1] Cf. O. W. Richardson, " Camb. Phil. Proc.," Vol. XII, p. 144 (1902); " Phil. Mag.," Vol. X, p. 242 (1905).

CHEMICAL ACTION ON SOLIDS.

When the ionization results from the chemical action of a gas on a solid the problem is more complicated than that furnished by a purely gaseous system. As an illustration we may consider the reaction

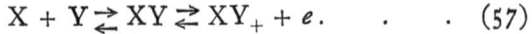

$$X + Y \rightleftarrows XY \rightleftarrows XY_+ + e. \qquad . \qquad . \quad (57)$$

where X and XY are solids and Y is a gas. Consider the equilibrium between the products inside a cavity in the solid X which contains a certain amount of the gas Y. Let the equilibrium concentrations of X, Y, XY_+ and e be C_1, C_2, C_3, and C_4 respectively. Then from (52)

$$C_4 = \frac{C_2 C_1}{C_3} e^{\int \frac{q}{RT^2} dT} \qquad . \qquad . \quad (58)$$

where q is the heat of the reaction at constant volume per molecule, when the reaction takes place in the gaseous phase. Now C_1 and C_3 are the molecular concentrations of the saturated vapours of the corresponding solids and are of the form

$$A e^{\int \frac{L}{RT^2} dT}$$

where L is the appropriate latent heat of evaporation. Thus if Q is the heat of reaction calculated on the assumption that the products present in the solid phase are decomposed and formed as solids

$$C_4 = A C_2 e^{\int \frac{Q}{RT^2} dT} \qquad . \qquad . \quad (59)$$

where A is independent of T. Thus if the concentration C_2 of the reacting gas is kept constant the concentration of the electrons in equilibrium in the cavity will vary with T, since Q does not vary much, in much the same way as in the purely gaseous case already considered, and as in the case of the purely thermionic emission.

Just as in dealing with the thermionic emission, we are not able to measure the equilibrium concentration of the electrons but only the rate at which they are emitted by the solid surface in presence of the gas. If the state of the surface is kept in the same condition as in the state of equilibrium it will emit in unit time approximately as many electrons as return to it under the equilibrium conditions, the approximation arising from the fact that here we are neglecting electron reflexion. It follows from the kinetic theory of gases that, to this degree of accuracy, the saturation current will be proportional to $C_4T^{\frac{1}{2}}$; which, combined with (59), shows that under these conditions a formula of the type $i = AT^{\frac{1}{2}}e^{-b/T}$ will be close to the truth.

It is possible, however, in experiments on the emission of electrons by the action of gases on solids to arrange matters so that the state of the solid is a long way from that which corresponds to the condition of equilibrium. Thus in Haber and Just's[1] experiments with metallic liquids the surface of the metal is renewed as quickly as it is attacked; so that the conditions correspond to the commencement of the reaction rather than to a state of equilibrium. Under these circumstances the saturation current will measure the initial velocity of the reaction defined by the left-hand side of (57). Thermodynamics is inadequate to determine the relation between the velocity of such a reaction and the temperature; but it has been found empirically that, in all cases of chemical reaction which are sufficiently simple to afford any analogy with the type now under consideration, the velocity is very closely proportional to $e^{-b/T}$, where b is constant.

One conclusion at least emerges clearly from this discussion, and that is, that the fact that the thermionic currents satisfy the formula $i = AT^{\frac{1}{2}}e^{-b/T}$ affords no evidence either for or against the view that these currents originate from chemical action. To settle this question it is necessary to appeal to evidence of a different character.

We shall now postpone the further development of the theory of the emission of electrons from hot bodies until some of the experimental results bearing on the conclusions already reached have been considered.

---

[1] See p. 308.

CHAPTER III.

TEMPERATURE VARIATION OF ELECTRONIC EMISSION.

THE first experiments on this subject were made by the writer [1] in order to test the theory developed on p. 35 of the last chapter. The elements investigated were platinum, carbon, and sodium. Since then measurements of the total emission at different temperatures have been made by H. A. Wilson [2] and by the writer [3] on platinum in atmospheres of hydrogen and other gases, by Wehnelt [4] on different metallic oxides, by G. Owen [5] on the Nernst filament, by Deininger [6] on platinum, carbon, tantalum, and nickel, in each case with the element alone and also when covered with a layer of lime, by Horton [7] on platinum covered with calcium and with lime, by Martyn [8] on platinum covered with lime in atmospheres of air and of hydrogen, by Jentztsch [9] on most metallic oxides, by Fredenhagen [10] on the emission from sodium and potassium, by Pring and Parker [11] and by Pring [12] on carbon, by Langmuir [13] on tungsten, tantalum, molybdenum, platinum, thorium

[1] "Camb. Phil. Proc.," Vol. XI, p. 286 (1901); "Phil. Trans., A.," Vol. CCI, p. 497 (1903).

[2] "Phil. Trans., A.," Vol. CCII, p. 243 (1903); Vol. CCVIII, p. 247 (1908).

[3] *Ibid.*, Vol. CCVII, p. 1 (1906); "Jahrb. d. Rad. u. Elektronik," Vol. I, p. 300 (1904).

[4] "Sitzungsber. der physik. med. Soc. Erlangen," p. 150 (1903); "Ann. der Phys.," Vol. XIV, p. 425 (1904); "Phil. Mag.," Vol. X, p. 88 (1905).

[5] "Phil. Mag.," Vol. VIII, p. 230 (1904).

[6] "Ann. der Physik," Vol. XXV, p. 285 (1908).

[7] "Phil. Trans., A.," Vol. CCVII, p. 149 (1907).

[8] "Phil. Mag.," Vol. XIV, p. 306 (1907).

[9] "Ann. der Phys.," Vol. XXVII, p. 129 (1908).

[10] "Verh. der Deutsch. Physik. Ges.," Jahrg. 14, p. 386 (1912).

[11] "Phil. Mag.," Vol. XXIII, p. 192 (1912).

[12] "Roy. Soc. Proc., A.," Vol. LXXXIX, p. 344 (1914).

[13] "Phys. Rev.," Vol. II, p. 450 (1913); "Trans. Amer. Electrochem. Soc.," p. 354 (1916).

and carbon, by the writer[5] and by K. K. Smith[1] on tungsten, by Schlichter[2] on platinum and nickel, by Stoekle[3] on molybdenum, and by Dushman[4] on titanium and iron. The work of Langmuir and Smith is characterized by extreme care in the elimination of gaseous impurities and exemplifies the most recent advances in technique.[5] The researches above mentioned are of a quantitative and, for the most part, extended character. In addition a number of investigations dealing with special points will be referred to in the sequel. Although the authors of the researches just enumerated differ considerably in the final interpretation of their experimental results, they are in agreement as to the general character of the variation of the rate of emission of the electrons by various hot bodies with temperature. In all cases it has been found that if the material experimented on is in a condition which does not change with lapse of time the rate of emission of electrons increases with enormous rapidity as the temperature is raised. This is true whether the substance under investigation is in a good vacuum or is surrounded by various gases. The extreme rapidity of this variation is well shown in Fig. 8 which represents the results of the writer's early experiments with sodium. The observations recorded extend over a range of temperature from $217°$ C. to $427°$ C. whilst the corresponding currents increased from $1·8 \times 10^{-9}$ amp. to $1·4 \times 10^{-2}$ amp. Thus with a rise of temperature of a little over $200°$ C. the current increases by a factor of 10^7. In order conveniently to exhibit all the values on the same diagram the curve is shown by means of a number of branches, in each of which, proceeding from left to right, the scale of the ordinates is successively reduced by a factor of 10. Thus, starting from the left-hand side, in the first curve the unit of current is 10^{-9} amp., in the second 10^{-8}, and so on. The various crosses which lie vertically over one another represent the same

[1] " Phil. Mag.," Vol. XXIX, p. 811 (1915).
[2] " Ann. der Phys.," Vol. XLVII, p. 573 (1915).
[3] " Phys. Rev.," Vol. VIII, p. 534 (1916), but cf. *ibid.*, Vol. IX, p. 500 (1917).
[4] Cf. Langmuir, " Trans. Amer. Electrochem. Soc.," p. 534 (1916).
[5] Cf. also O. W. Richardson, " Phil. Mag.," Vol. XXVI, p. 345 (1913), and Stoekle., loc. cit.

observation on different scales.   It will be noticed that the successive branches are very similar to one another ; so that the general character of the temperature variation is much the same at all temperatures.   As the temperature is reduced the current continuously approaches the value zero but never actually reaches it.   The experiments to which Fig. 8 refers were probably affected to some extent by the presence of a surrounding gaseous atmosphere, but however carefully gaseous contamination has been avoided, it has always been found that

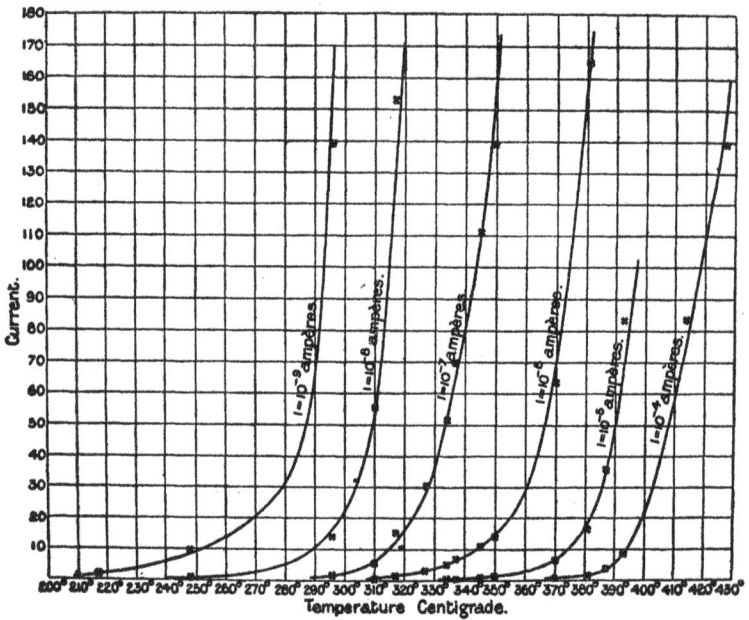

Fig. 8.

the general character of the temperature variation is of the kind shown in the figure.   The difference between different substances lies in the temperature at which the emission becomes appreciable ; and this temperature determines the whole scale of the diagram.   With most substances the currents cannot be measured on a sensitive galvanometer at temperatures below 1000° C.   A correspondingly larger interval of temperature is then required in order to change the current in a given proportion.

   This will become clearer if we consider the matter from a

more quantitative standpoint. In the last chapter two principal formulæ were developed for the saturation current i from unit area of a hot body at temperature T°K, viz. :—

$$i = AT^{\frac{1}{2}}e^{-b/T} . \qquad . \qquad . \qquad . \qquad (1)$$

and

$$i = CT^2 e^{-d/T} . \qquad . \qquad . \qquad . \qquad (2)$$

where A, b, C, and d are constants. Obviously these formulæ cannot both be true. As a matter of fact both are approximations and (2) rests on a more solid theoretical basis than

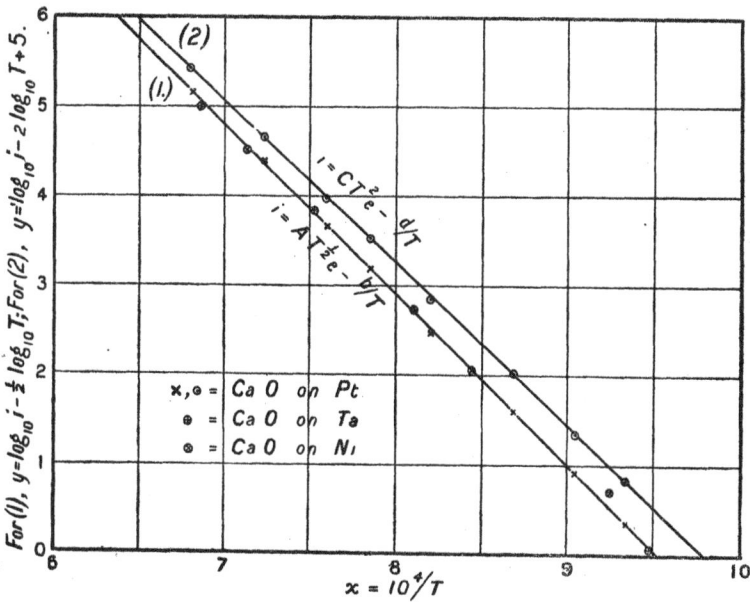

Fig. 9.

(1). According to the theory of Chapter II, equation (2) should be a very close approximation to the truth. In order to test the relative merits of the equations we may take logarithms of both sides, obtaining from (1),

$$\log_{10} i - \tfrac{1}{2} \log_{10} T = \log_{10} A - b/2 \cdot 303\,T \qquad . \qquad (3)$$

and from (2),

$$\log_{10} i - 2 \log_{10} T = \log_{10} C - d/2 \cdot 303\,T \qquad . \qquad (4)$$

According to (1) we should get a straight line by plotting $\log_{10} i - \tfrac{1}{2} \log_{10} T$ against T^{-1} and according to (2) the same result should follow if we plot $\log_{10} i - 2 \log_{10} T$ against T^{-1}.

In Fig. 9 some very consistent observations by Deininger,[1] on the emission from lime coated platinum wires, are treated in this way, the saturation current per unit area being given in electrostatic units. The points shown thus: × represent the values of $\log_{10} C - \frac{1}{2} \log_{10} T$, and the points thus: ⊙ the values of $5 + \log_{10} C - 2 \log_{10} T$, each plotted against $10^4/T$. The points marked ⊕ and × may be left out of consideration for the present. The points shown represent a variation of C from 75 e.s.u. to approximately 3×10^6 e.s.u., the corresponding range of temperature being from 798° C. to 1198° C. Both sets of points are seen to fall on straight lines almost as exactly as they can be drawn; so that if we regard (1) and (2) as empirical formulæ there is nothing to choose between them. Each is capable of expressing the experimental results with the exactness required by the accuracy of the measurements. It is clear that the values of A and b or C and d may be deduced from the intercepts on the axes of lines like (1) or (2) respectively in Fig. 9. As the experiments are unable to decide between the relative accuracies of the two formulæ, and as the formula (1) was the first in the field and usually occurs in the literature, we shall make most frequent use of it in this book, although (2) is more satisfactory from a theoretical standpoint. After all, the difference between the two equations may be considered to lie in the interpretation of the quantities A, b, C, and d, and if A and b are given for a particular substance and temperature, C and d may be obtained from the relations

$$C = A e^{-\frac{1}{2}} T^{-\frac{1}{2}} \qquad . \qquad . \qquad . \qquad (5)$$

and
$$d = b - \frac{1}{2} T \qquad . \qquad . \qquad . \qquad (6)$$

which are valid to the degree of accuracy within which (1) and (2) can be regarded as consistent.

The experimental evidence shows that the validity of equation (2) (or (1)) is perfectly general and covers the very large number of substances investigated over the whole range of experimentation, provided no permanent change in the composition of the surface of the substance concerned takes

---

[1] Loc. cit., p. 296.

place as a result of the treatment. The only known exception appears to be that with certain specimens of commercial osmium filament the writer and H. L. Cooke[1] observed that the curve corresponding to Fig. 9 consisted of two straight lines meeting at an angle. It is probable that this is due to some reversible, possibly allotropic, change in the structure of the material. The range of thermionic current over which the formula has been found to apply is in many cases very large indeed. Thus K. K. Smith[2] has confirmed it in the case of tungsten over a range of temperature such that the thermionic current varied by a factor of 10^{11}. His results also show that even over this extended range there is nothing to choose between equations (1) and (2), both of which express the experimental results with exactness. Such consistency is, of course, only obtainable after the material has been thoroughly freed from impurities. New filaments frequently give data yielding exceptional values of the constants A and b.

Conditions Affecting the Attainment of Saturation.

In making measurements of the number of electrons emitted from hot bodies it is essential that the currents should be saturated, otherwise only part of the electrons emitted by the hot body will reach the electrode and the measured values will be too small. It is therefore necessary that the applied potential difference between the emitting substance and the receiving electrode should be at least as great as the smallest potential difference required to cause saturation. Generally speaking, the applied potential difference may have any value greater than this, provided the experiments are made in a good vacuum. If, however, a gaseous atmosphere is present we have seen in Chapter I that, if the potential gradient exceeds a certain value depending on the nature and pressure of the gas, ionization by collision will occur and the measured currents will be larger than those due to the unassisted electronic emission. When possible this difficulty should be

---

[1] " Phil. Mag.," Vol. XXI, p. 408 (1911).
[2] *Ibid.*, Vol. XXIX, p. 802 (1915).

avoided by making the experiments under the best attainable
vacuum conditions, as it is difficult to make exact allowance
for the effect of ionization by collisions in experiments of
this character.

The type of curve connecting the current and applied potential difference which is most frequently obtained under fairly
good vacuum conditions is shown in Fig. 10. This represents
observations made by the writer [1] with a U-shaped carbon
filament surrounded by a cylindrical electrode, the pressure
being 0·003 mm. It will be seen that approximate saturation
is attained at about 30 volts, although there is a further rise
of about 10 per cent of the total value on increasing the volt-

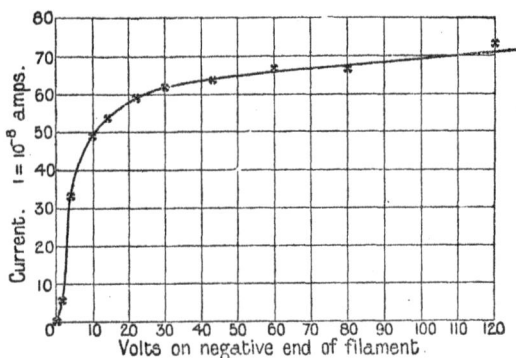

Fig. 10.

age to 120. This further increase [2] is almost always, as in Fig.
10, proportional to the increase in the applied potential. It
usually diminishes with continued use of a given tube, and it
appears to be due either to the evolution of gas from the hot
filament or to the presence of a layer of condensed gas on the
electrode, or to both these circumstances. The writer has not
observed this effect in a tube which has been well glowed out
and exhausted in the vacuum furnace before testing, although
a case of its appearance under these conditions has been recorded by K. K. Smith.[3] As an example of the extent to

[1] " Phil. Trans., A.," Vol. CCI, p. 520 (1903).
[2] Some recent observations of this phenomenon have been recorded by
H. Lester, " Phil. Mag.," Vol. XXXI, p. 549 (1916).
[3] Loc. cit.

which complete saturation is attainable in experiments of this character the following figures given by Deininger[1] may be cited :—

Potential Difference (volts)—

0	3	5	10	12	15	20	25	30

Current ($1 = 10^{-8}$ amp.)—

0·0	23·7	54·7	122·3	167·9	186·2	202·6	206·2	208·1

P.D.→

40	50	60	70	80	90	100	110	120	130	140	180

C.→

209·0	209·0	209·0	211·7	208·1	204·4	204·4	208·1	208·1	208·1	208·1	208·1

In this experiment the temperature must have been about the same as in the case to which Fig. 10 refers. The pressure is given as less than 0·001 mm. It will be seen that there is here no perceptible increase in the current between 25 and 180 volts.

It is only when the thermionic current densities are comparatively small that saturation is attainable with potential differences of 20 volts or under. With the large emissions which occur at very high temperatures the potential difference required for saturation may be very much higher owing to the mutual repulsion of the emitted electrons. In fact with a given difference of potential between the electrodes a stage is reached when further increase in the temperature causes no increase in the currents flowing. This effect was first observed by Lilienfeld[2] who explained it in general terms[3] as due to the effect of the volume charge of the emitted electrons in opposing the transportation of the current. These phenomena are of considerable technical importance and have been investigated very fully by Langmuir,[4] whose measurements of the thermionic current between two hairpin-shaped tungsten filaments, under fixed differences of potential, at various temperatures, are exhibited in Fig 11. The experimental values lie along the broken curves, the continuous curve representing the saturation current at different temperatures.

[1] "Ann. der Phys.," Vol. XXV, p. 294 (1908).

[2] *Ibid.*, Vol. XXXII, p. 673 (1910); "Leipziger Ber," Vol. LXIII, p. 534 (1911).

[3] "Phys. Zeits," IX, p. 193 (1908).

[4] "Phys. Rev.," Vol. II, p. 453 (1913).

It will be noticed that the broken curves coincide with the continuous curve up to a certain temperature which is lower the smaller the applied potential difference. Beyond this point the currents are below the saturation value, but they still increase with increasing temperatures of the filament. At still higher temperatures the broken curves bend round, and the current under a given potential difference becomes entirely

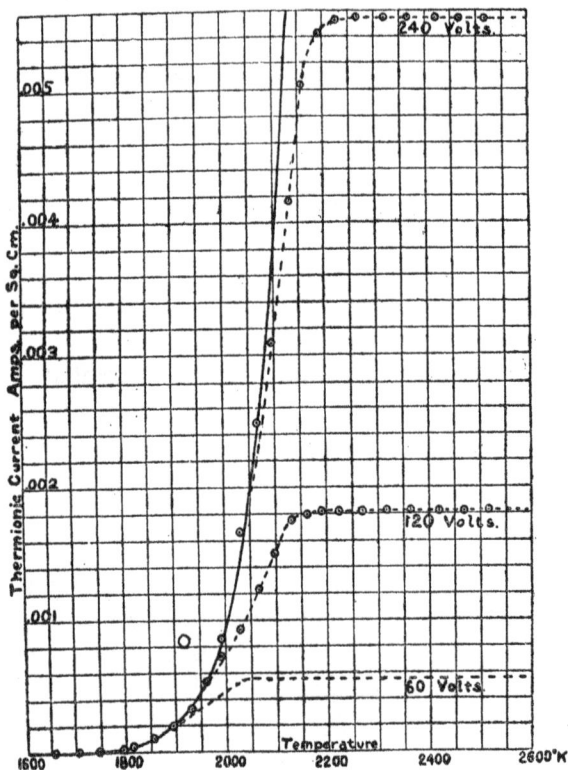

FIG. 11.

independent of the temperature of the hot filament. Owing to the mutual repulsion of the electrons a given difference of potential applied over given boundaries is only capable of forcing a definite number of electrons across the intervening space in unit time, no matter how many may be available.

In order more clearly to see how this comes about let us consider the simplest possible case, that of an emitting plane opposite a parallel conducting plane which represents the receiving electrode. Take the axis of x perpendicular to the planes and let V be the electrostatic potential at any point. Then the value of V is governed by Poisson's equation which reduces to

$$\frac{d^2V}{dx^2} = -4\pi\rho \quad . \quad . \quad . \quad . \quad (7)$$

ρ being the volume density of electricity at any point. When there are no electrons emitted ρ is zero at every point; so that $\frac{dV}{dx}$ is constant and the relation between V and x may

be represented by the straight line PQT in Fig. 12. When the plate is emitting electrons these will give rise to a negative value of ρ in the space as they cross it, the curve between V and x will therefore everywhere be convex downwards and if the number is small, the graph of V will be similar to PST. As the emission and $-\rho$ increase this curve will sink below PQT until the stage PRT is reached, where the tangent at P is horizontal. Any further

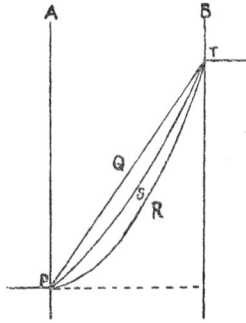

FIG. 12.

increase in the supply of electrons will now have no effect on the distribution of potential between the plates because there is nothing to drag them away from AP. We here assume that the emission velocities of the electrons are zero.

Now let us see how the current will depend on the applied potential when the condition that $\frac{dV}{dx}$ vanishes at the hot plate is satisfied. Let $+ e_1$ denote the numerical value of the negative charge of an electron and $+ \rho_1$ the numerical value of the negative density of charge at any point. An electron at a point where the potential is V (if V = 0 at the hot plate) will have acquired an amount of kinetic energy given by the equation

$$\tfrac{1}{2}mv^2 = Ve_1 \quad . \quad . \quad . \quad . \quad (8)$$

The current per unit area carried by the electrons at that

point will be

$$i = v\rho_1 \quad . \quad . \quad . \quad . \quad (9)$$

and equation (7) may be written

$$\frac{d^2V}{dx^2} = 4\pi\rho_1 \quad . \quad . \quad . \quad . \quad (10)$$

By elimination of ρ and v from (8), (9), and (10)

$$\frac{d^2V}{dx^2} = 2\pi i \sqrt{\frac{2m}{e_1 V}}$$

and integrating, subject to $\dfrac{dV}{dx} = 0$ when $V = 0$,

$$\left(\frac{dV}{dx}\right)^2 = 8\pi i \sqrt{\frac{2mV}{e_1}} \quad . \quad . \quad . \quad (11)$$

By integrating again, subject to $V = 0$ when $x = 0$, and solving for i we find

$$i = \frac{\sqrt{2}}{9\pi} \sqrt{\frac{e_1}{m}} \frac{V^{3/2}}{x^2} \quad . \quad . \quad . \quad (12)$$

This calculation is a development of one first given by J. J. Thomson [1] and was first carried out in the form here given by C. D. Child.[2]

The same general method of treatment with the appropriate modification of Poisson's equation has been applied to the case of a circular wire surrounded by a coaxial cylinder by Langmuir who finds

$$i = \frac{2\sqrt{2}}{9} \cdot \sqrt{\frac{e_1}{m}} \frac{V^{3/2}}{r\beta^2} \quad . \quad . \quad . \quad (13)$$

where r is the radius of the cylinder, and

$$\beta = \log\frac{r}{a} - \overset{*}{\frac{2}{5}}\left(\log\frac{r}{a}\right)^2 + \frac{11}{120}\cdot\left(\log\frac{r}{a}\right)^3$$

$$- \frac{47}{3300}\left(\log\frac{r}{a}\right)^4 + \ldots,$$

a being the radius of the wire. He has suggested that experi-

---

[1] "Conduction of Electricity through Gases," 1903 edition, p. 187; 1906 edition, p. 223.

[2] "Phys. Rev.," Vol. XXXII, p. 498 (1911).

* I am informed by Professor Fleming that this factor should be $\frac{2}{7}$ not $\frac{2}{5}$.

ments with concentric cylinders using equation (13) could be made so as to determine the value of e_1/m with great precision.[1]

Langmuir[2] has argued from a consideration of the dimensions of the equations that, whatever may be the geometrical relation between the emitting and receiving electrode, the current in the infra saturation region will be equal to $V^{3/2}$ multiplied by a factor depending on e_1/m and the geometry of the system. For this argument to be valid it is necessary that the velocity and density of the electrons should be capable of being varied in an arbitrary manner throughout every point of the space right up to the cathode. The existence of finite emission velocities independent of the field intensities prevents this condition from being satisfied. The thesis that the current varies as $V^{3/2}$ cannot therefore be considered to be established in general. It is, however, found in practice that the currents can be expressed with fair approximation as being proportional to $V^{3/2}$ for a considerable variety of two electrode structures and over a range of voltage varying with the structure, as, for example, over the flat parts of Fig. 11.

There are considerable deviations when the applied voltages are small and also at high voltages as the saturation value of the current is being approached.

The deviations at low voltages are of a fundamental character. In order to understand them it is necessary to consider critically two of the assumptions which have been made in the foregoing calculations, namely, that the initial velocities are zero, and that the vanishing of the potential gradient at the cathode is an infra-saturation condition. The validity of the latter assumption certainly seems doubtful because we can argue with at least equal plausibility that it is the condition which should hold when saturation is just attained. We have seen that inasmuch as the volume density ρ is everywhere negative the V, x curve is always concave upwards. If, therefore, it is horizontal at the cathode (Fig. 12), the electric force in the space will at every point be urging the electrons towards the anode. Any electron which escapes from the cathode will

---

[1] Cf. S. Dushman, "Phys. Rev.," Vol. IV, p. 121 (1914).
[2] Loc. cit.

thus reach the anode and the condition is one for the attainment of saturation. If the force at the cathode does not vanish but is so directed as to draw the electrons towards the anode the condition is equally one for saturation. If it is oppositely directed, no matter how small it may be, it will stop the current provided the initial velocities are *zero*. It appears then that if the initial velocities are zero there is an instability in the current voltage phenomena, the current jumping suddenly from zero to the saturation value in passing through that voltage for which $\dfrac{dV}{dx}$ vanishes at the cathode. But this is quite a different picture of the phenomena from that given by the calculations of Child and Langmuir which are based on the same hypotheses, and we are driven to inquire how it is that those calculations show as substantial a measure of agreement with the facts as they do.

As a matter of fact the initial velocities are not zero. We shall see in Chapter V. that the electrons are emitted with a complicated distribution of velocities whose mean value is comparable with that which an electron would acquire in falling through a difference of potential of 0·25 volt. Actually the vanishing of the potential gradient at the cathode is the condition for incipient saturation. When the current is below the saturation value the potential gradient at the cathode is finite, and so directed as to return the emitted electrons to it. Since the negative volume density necessitates the V, x curve being uniformly concave upwards, it follows that in the infrasaturation region there is a surface over which the potential has a minimum value located in the space between the cathode and the anode. It is the initial velocity of the electrons which carries them, or rather, such of them as have sufficiently high initial velocities, through the opposing field in the neighbourhood of the cathode. The surface of minimum potential is also the surface at which the potential gradient vanishes, and the electrons which get through it pass to the anode under an accelerating field. When the currents are a very long way from saturation the surface of minimum potential is close to the anode, and as the currents increase it moves towards the cathode with which it becomes coincident as saturation sets in. Thus the vanishing of the potential gradient at the cathode is actually a saturation condition. The reason why the currents are found in practice to vary very approximately as the 3/2 power of the voltage over a very considerable range must be that in favourable cases where the electron emission is large there is a considerable range of current over which the surface of minimum potential is very close to that of the cathode, and the minimum potential is not much below the potential of the cathode.

It will be seen that the surface of minimum potential divides the space between the cathode and anode into two portions in which the behaviour of the electrons is governed by quite different considerations. In the positive part of this space, that is, on the side towards the anode, the electrons are moving in an accelerating field, and the calculations of Child and Langmuir will apply, subject to a correction for the initial velocities, if the potentials and distances are measured from the surface of minimum potential instead of from the cathode. In the negative part of the space the electrons are in a retarding field and the conditions are mainly similar to those considered on pp. 173 *et seq*. The conditions are not, however, those of a true equilibrium, as there is a steady flow of electrons towards the anode. This makes the resulting differential equations difficult, and so far as I am aware no one has yet succeeded in obtaining an exact solution. In the case of parallel planes, however, Schottky[1] has succeeded in deducing the following equation for the current i at different potentials (V) :—

$$i = \frac{\sqrt{2}}{9\pi} \sqrt{\frac{e}{m}} \left(\frac{k\mathrm{T}}{\pi e}\right)^{3/2} (\sqrt{n} - 1)(\sqrt{n} + 2)^2 \Big/ (l - d)^2 \qquad (13\cdot1)$$

where
$$n = 1 + \pi \log \frac{s}{i} + \frac{\pi e}{k\mathrm{T}}\mathrm{V} \qquad . \qquad . \quad (13\cdot2)$$

$$d = \frac{\sqrt{\frac{m}{ei}}\left(2\frac{k\mathrm{T}}{\pi m}\right)^{3/4}}{3 - \frac{i}{s\sqrt{1 + \pi \log s/i}}} \tan^{-1}\sqrt{\frac{s}{i} - 1} \quad . \quad (13\cdot3)$$

[1] " Phys. Zeits.," Vol. XV, p. 526 (1914).

is the distance of the surface of minimum potential from the cathode, k is Boltzmann's constant, T is the characteristic temperature of the electrons, s is the saturation current, and d the distance between the electrodes. These formulæ appear to be exact enough for all practical purposes. In the case of currents saturating with about 30 volts they indicate errors varying between 10 and 40 per cent, when equation (12) is used to calculate the currents under a given voltage in the infra-saturation region. In the case of cylindrical electrodes the calculations are still more difficult, and it has only been found possible to determine the minimum potential within rather wide limits.[1] These corrections would be of importance, for example, in using equation (13) to determine e/m with accuracy.

The effect of the mutual repulsion of the electrons in preventing the attainment of saturation will be important only when the saturation currents are of considerable magnitude ; with sufficiently small currents this effect will vanish. At relatively low temperatures when the emission is comparatively small, if the hot body is surrounded by the receiving electrode, we should expect saturation to be attained without the application of any potential difference ; since all the electrons are emitted with some velocity, and any velocity, however small, will be sufficient to carry them across to the electrode ultimately. This supposition does not accord with the facts as observed in experiments with electrically heated wires surrounded by coaxial cylindrical electrodes. Under these conditions it is the writer's experience that potential differences comparable with two or three volts are required to cause saturation even when dealing with the smallest currents which are convenient to measure. There can be very little doubt that in these cases one cause operating against the attainment of saturation is the effect of the magnetic field, due to the heating current, on the motion of the electrons.

To see how this comes about consider the case of a hot wire, of circular section and radius a, surrounded by a coaxial

---

[1] Schottky, "Verh. der deutsch. Physik. Ges.," Vol. XVI, p. 490 (1914).

cylindrical electrode of radius b. Let V_1 be the potential difference in volts between the wire and the cylinder. The electric intensity R is everywhere radial and at distance r from the axis is given in electro-magnetic units by

$$R = V_1 \times 10^8/r \log b/a = A/r \qquad . \qquad . \quad (14)$$

The magnetic intensity H lies in circles about the axis of the wire, and, if j is the current in amperes, its value at distance r is

$$H = 2j/10r = B/r \qquad . \qquad . \qquad . \quad (15)$$

On account of its direction the magnetic field will not affect the angular velocity of the electrons about the axis. Disregarding this rotation, the paths of the electrons are periodic curves, in the plane containing the axis, which keep intersecting the surface of the emitting cylinder. The effect of the neglected rotation is simply to convert these plane curves into spirals about the axis. There is a certain maximum distance, under given conditions, which an electron is able to travel from the axis, and unless this is equal to or greater than the radius of the outer cylinder the electrons will return to the surface of emission and will contribute nothing to the thermionic current.

If r is the perpendicular distance of an electron from the axis, z its distance from a fixed plane perpendicular to the axis, and θ the angle the plane containing the axis and the electron makes with a fixed plane through the axis, the equations of motion of the electron are :—

$$m\frac{\partial^2 r}{\partial t^2} - mr\left(\frac{\partial \theta}{\partial t}\right)^2 = Re - He\frac{\partial z}{\partial t} = \frac{Ae}{r} - \frac{Be}{r}\frac{\partial z}{\partial t} \quad . \quad (16)$$

$$m\frac{\partial^2 z}{\partial t^2} \qquad = He\frac{\partial r}{\partial t} \qquad = \frac{Be}{r}\frac{\partial r}{\partial t} \qquad . \quad . \quad (17)$$

$$\frac{\partial}{\partial t}\left(r^2\frac{\partial \theta}{\partial t}\right) \qquad = 0 \qquad . \qquad . \qquad . \qquad . \qquad . \quad (18)$$

From (18)

$$\frac{\partial \theta}{\partial t} = \frac{a^2}{r^2}\dot{\theta}_0 \qquad . \qquad . \qquad . \qquad . \quad (19)$$

if $\qquad \dfrac{\partial \theta}{\partial t} = \dot\theta_0$ when $r = a$.

From (17)

$$\frac{\partial z}{\partial t} = \dot z_0 + \frac{Be}{m} \log \frac{r}{a} \qquad . \qquad . \qquad . \quad (20)$$

if $\qquad \dfrac{\partial z}{\partial t} = \dot z_0$ when $r = a$.

Substituting these values of $\dfrac{\partial \theta}{\partial t}$ and $\dfrac{\partial z}{\partial t}$ in (16) and integrating subject to

$$\frac{\partial r}{\partial t} = \dot r_0 \text{ when } r = a,$$

$$\frac{\partial r}{\partial t} = \left[\dot r_0{}^2 + a^2 \dot\theta_0{}^2 \left(1 - \frac{a^2}{r^2} \right) - \frac{B^2 e^2}{m^2} \left(\log \frac{r}{a} \right)^2 \right.$$
$$\left. + \frac{2(Ae - Be\dot z_0)}{m} \log \frac{r}{a} \right]^{1/2} . \qquad . \quad (21)$$

The maximum value r_m of r is given by $\dfrac{\partial r}{\partial t} = 0$ or from (21), after substituting the values of A and B in terms of V_1 and j, by

$$V_1 = 10^{-8} \log \frac{b}{a} \left[\frac{j^2 e}{50 m} \log \frac{r_m}{a} + \frac{j \dot z_0}{5} \right.$$
$$\left. - \frac{m \dot r_0{}^2 + m a^2 \dot\theta_0{}^2 \left(1 - \dfrac{a^2}{r_m{}^2} \right)}{2e \log r_m/a} \right] . \qquad . \quad (22)$$

Thus if r_m is to be just equal to the radius b of the outer cylinder,

$$V_1 = 10^{-8} \log \frac{a}{b} \left\{ \frac{j^2 e}{50 m} \log \frac{b}{a} + \frac{j}{5} \dot z_0 \right.$$
$$\left. - \frac{m \dot r_0{}^2 + m a \theta^2{}_0{}^2 \left(1 - \dfrac{a^2}{b^2} \right)}{2e \log b/a} \right\} . \qquad . \quad (23)$$

If V_1 has a value equal to or greater than this the electrons will reach the electrode and form part of the current, otherwise they will not do so. If j is very small the right-hand side of (23) is negative, indicating that the current will be able to flow against an opposing potential owing to the initial emission velocities. The point of immediate interest, however, is the first term, on the right-hand side of (23), which is always positive and is independent of the emission velocities. This shows that owing to the action of the magnetic field due to the heating current, a definite potential is necessary in order to drag the electrons across to the electrode. With thin wires, which require only a small current to heat them, this potential difference is unimportant. Thus if $b/a = 200$ and $j = 1$ ampere, V_1 is only about 0·2 volt. On the other hand, with thick wires, which require large heating currents, the necessary values of V_1 may be quite large.

Another important factor which has to be taken into account, especially with thin wires, is the drop of potential along the wire due to the flow of the heating current. This is usually comparable with 1 volt per cm. In order to ensure that no part of the wire is at a positive potential compared with the cylinder, it is necessary that the *positive* end of the wire should be at a potential at least as low as that of the cylinder. If the potentials are applied at the negative end of the wire, it will appear from this cause alone that an additional negative potential equal to the fall along the hot wire has to be applied in order to ensure complete saturation. Where the conditions are such that the current would be varying as the 3/2 power of the voltage if the cathode were an equipotential surface the effect of the fall of potential due to the current flow can be allowed for by the following method. Let V_1 be the potential difference between the positive end of the cathode filament and the anode, let V_0 be the potential drop in the filament whose length is l. Let the current per unit length in general be $CV^{3/2}$ then the potential at a point distant x from the positive end of the filament is $V_1 + \frac{x}{l}V_0$ and the current from a length dx at this point is

$$C\left(V_1 + \frac{x}{l}V_0\right)^{3/2} dx$$

and the total current is

$$C\int_0^l \left(V_1 + \frac{x}{l}V_0\right)^{3/2} dx = \frac{2}{5}C\frac{l}{V_0}\{(V_1 + V_0)^{5/2} - V_1^{5/2}\}$$

That the mutual repulsion of the electrons, the magnetic field due to the heating current, and the drop of potential along the wire, also due to the last-named cause, are the chief general factors which prevent the attainment of saturation is strikingly shown by some recent experiments by Schottky.[1] Using concentric cylinders the thermionic currents with small differences of potential, both accelerating and retarding, were measured under conditions such that the heating current was cut out at the instant of measurement. By means of a suitable in-and-out switch,[2] operating continuously, matters were arranged so that no appreciable variation in the temperature of the wire ensued thereby. Under these conditions the drop of potential along the filament and the magnetic field due to the heating current are eliminated, and it was found that the current saturated at zero potential difference; except for an effect, which was smaller the lower the temperature and smaller the current, arising from the mutual repulsion of the emitted electrons. In determining the actual difference of potential between the wire and the electrode, it was found necessary to add to the apparent applied potential difference registered by the voltmeter, a difference of potential equal to the contact potential between the two metals used. Thus experiments of this kind can be used to measure contact differences of potential under good vacuum conditions without displacing the surfaces subject to test.

Schlichter,[3] by using methods of heating the electrode which do not involve the passage of an electric current through it, has been able to show that the electron current with 220 volts driving potential is only about 10-20 per cent greater than that under zero potential difference, when the hot metal has been thoroughly glowed out. The special conditions which affect the attainment of saturation in the case of freshly heated metals will be referred to again on page 199, Chap. VI.

[1] " Ann. der Phys.," Vol. XLIV, p. 1011 (1914).
[2] Cf. v. Baeyer, " Verh. der Deutsch. Physik. Ges.," Vol. X, p. 98 (1908).
[3] " Ann. der Phys.," Vol. XLVII, p. 573 (1915).

INFRA SATURATION CURRENTS IN PRESENCE OF GASES.

In general the relation between currents and voltage when gases are present is very complicated, depending on the nature and pressure of the gas, the electron emission from the cathode, the applied potential difference, and the geometry of the electrodes. For each gas it is found that there is a definite potential difference, called the ionization potential, such that unless the electrons acquire an amount of kinetic energy equivalent to that arising from a fall through this difference of potential they are unable to ionize the gas molecules by impact. Provided the potential difference between the electrodes is less than the ionization potential, no positive ions will form and the current will be carried entirely by electrons or by negative ions. If the gas pressure is sufficiently low the magnitudes of the currents under given voltages will then be determined by the considerations developed on pp. 65-78. As soon as the ionization potential is exceeded some positive ions will begin to form. The positive ions are much more massive than the electrons and the most noticeable effect they exert is to reduce the volume density of the negative electric charge in the space between the electrodes. If the gas is one with light molecules, such as hydrogen, and the pressure is relatively low, the current will show a gradual increase over the high vacuum values as the voltage is raised. If, however, the gas is one with heavy molecules, such as mercury vapour, and the pressure is not too low, the sudden generation of positive ions may be almost sufficient to extinguish the space charge effect of the electrons and there may be a sudden jump in the current to a value approaching saturation. Such unstable phenomena in the current voltage relations in the presence of gases at a low pressure have frequently been noticed by observers in this field.[1]

When the pressure of the gas is high, for example, comparable with the atmospheric pressure, and the hot electrode generates ions of one sign only, the conditions which

[1] Cf. Richardson and Bazzoni, "Phil. Mag.," Vol. XXXII, p. 428 (1916); Hagenow, "Phys. Rev.," Vol. XIII, p. 415 (1919).

determine the flow of the current differ from those for a high vacuum in one important particular. The velocity of the ions or electrons now, instead of being proportional to the square root of the potential difference reckoned from the emitting electrode, is proportional at every point to the electric intensity. When this fact, together with the effect of the space charge, is taken into account J. J. Thomson[1] has shown that the current i per unit area between two parallel plane electrodes, one of which is a source of ions of one sign only, is given, when the currents are small compared with the saturation value, by the equation

$$i = \frac{9k\mathrm{V}^2}{32\pi l^3},$$

where k is the mobility of the ions, V is the applied potential difference, and l is the distance between the planes. In this case the current varies as the square of the applied potential difference.

The presumption that the velocity of the ions is proportional to the electric intensity implies that they lose most of their kinetic energy at each collision with a gas molecule. With a number of gases such as mercury vapour and helium this is far from being the case; in fact it appears that such loss of energy is inappreciable unless the ions or electrons have acquired a certain critical kinetic energy as, for example, by falling through a corresponding "critical" potential difference. When the modifications appropriate to such gases[2] are made in the equations the currents between parallel plates at potentials (V) below the critical potential are found to be given by the equation

$$i = \sqrt{\frac{e}{2m}} \frac{\lambda}{2\pi l^3} \mathrm{V}^{3/2}$$

where λ is the mean free path of the ion or electron in the gas and l is the distance between the plates. In this case it appears that the current varies as the voltage raised to the

[1] "Conduction of Electricity through Gases," 2nd edition, p. 209.
[2] Richardson and Bazzoni, "Phil. Mag.," Vol. XXXII, p. 433 (1916).

power 3/2. This result should be independent of the geometry of the electrodes provided the pressure is reasonably high and the currents are a long way from saturation.

The Values of the Constants.

A considerable number of the researches enumerated at the beginning of this chapter are of a sufficiently extensive character to enable the constants of the emission formula to be deduced from the measurements. The values for the elementary substances are given in the next table. The numbers given are the values of A_1, b, C, d, and ϕ_0, where $A_1 = A/\epsilon$,

$$\phi_0 = \frac{Rd}{\epsilon} \times 300$$ and A, b, C, and d, are the constants in equations (1) and (2), when i is expressed in electrostatic units. A_1 in fact is the constant in the equation

$$N = A_1 T^{\frac{1}{2}} e^{b/T} \qquad . \qquad . \qquad . \quad (24)$$

where N is the number of electrons emitted from unit area in unit time at temperature T. ϕ_0 is the potential difference in volts which is equivalent to the work an electron would have to do to escape from the substance, reduced to the absolute zero of temperature. In some cases the numbers have not been evaluated by the authors, in others obsolete values of the ionic charge ϵ have been used. I have reduced all the data to the common value $\epsilon = 4.8 \times 10^{-10}$ e.s.u :—

Material.	Observer.	A_1.	b.	C.	d.	ϕ_0(equivalent volts).
Carbon	1. [Richardson	10^{34}	7.8×10^4		7.55×10^4	6·48]
	2. Deininger	4.68×10^{25}	5.49×10^4	7.46×10^{10}	5.25×10^4	4·51
	3. Langmuir	1.49×10^{25}	4.87×10^4	1.78×10^{10}	4.57×10^4	3·92
	3·3 Langmuir		4.80×10^4			
Platinum	4. Richardson	7.5×10^{25}	4.93×10^4		4.75×10^4	4·1
	5. Wilson	6.9×10^{26}	6.55×10^4		6.3×10^4	5·45
	6. Wilson	1.17×10^{27}	7.25×10^4		7.0×10^4	6·0
	7. Richardson	5×10^{28}	6.78×10^4		6.55×10^4	5·65
	8. Deininger	3.06×10^{25}	6.1×10^4	4.9×10^{10}	5.85×10^4	5·02
	9. Horton	1.6×10^{25}	6.1×10^4		5.9×10^4	5·1
	10. [Wilson	2×10^{21}	2.8×10^4		2.56×10^4	2·18]
	11. Langmuir	2.02×10^{31}	8.0×10^4	2.42×10^{16}	7.7×10^4	6·62
	11a. Schlichter	7.2×10^{25}	5.11×10^4		4.9×10^4	4·2
Tungsten	12. Langmuir	1.55×10^{26}	5.25×10^4	1.86×10^{11}	4.95×10^4	4·25
	12a. K. K. Smith	3.0×10^{27}	5.47×10^4		5.20×10^4	4·46

6

Material.	Observer.	A_1.	b.	C.	d.	ϕ_0(equivalent volts).
Tantalum	13. Deininger	$2\cdot7 \times 10^{21}$	$4\cdot42 \times 10^4$	$4\cdot3 \times 10^6$	$4\cdot17 \times 10^4$	$3\cdot58$
	14. Langmuir	$7\cdot45 \times 10^{25}$	$5\cdot0 \times 10^4$	$8\cdot94 \times 10^{10}$	$4\cdot7 \times 10^4$	$4\cdot04$
Molybdenum	14·1 Stoekle	$6\cdot9 \times 10^{26}$	$5\cdot36 \times 10^4$			
Molybdenum	15. Langmuir	$1\cdot38 \times 10^{26}$	$5\cdot0 \times 10^4$	$1\cdot65 \times 10^{11}$	$4\cdot7 \times 10^4$	$4\cdot04$
Thorium	15·1 Langmuir	$1\cdot25 \times 10^{27}$	$3\cdot90 \times 10^4$			
Nickel	15a. Schlichter	$2\cdot9 \times 10^{25}$	$3\cdot4 \times 10^4$		$3\cdot3 \times 10^4$	$2\cdot9$
Iron	15·2 Dushman	$1\cdot5 \times 10^{22}?$	$3\cdot7 \times 10^4?$			
Titanium	15·3 Dushman	$8\cdot1 \times 10^{21}?$	$2\cdot8 \times 10^4?$			
Calcium	16. Horton	$1\cdot1 \times 10^{23}$	$3\cdot65 \times 10^4$		$3\cdot5 \times 10^4$	$3\cdot04$
Sodium	17. [Richardson	10^{31}	$3\cdot16 \times 10^4$		$3\cdot1 \times 10^4$	$2\cdot65$]

The data from which these numbers have been calculated are taken from the following list of papers. (The numbers are the numerals at the beginning of each row in the preceding table):—

Nos. 1, 4, and 17, "Phil. Trans., A.," Vol. CCI, p. 497 (1903). Nos. 2, 8, and 13, "Ann. der Phys.," Vol. XXV, p. 285 (1908). Nos. 5, 6, and 10, "Phil. Trans., A.," Vol. CCII, p. 243 (1903); Vol. CCVIII, p. 247 (1908). No. 7, "Phil. Trans., A.," Vol. CCVII, p. 1 (1906). Nos. 9 and 16, "Phil. Trans., A.," Vol. CCVII, p. 149 (1907). Nos. 11, 14, and 15, "Phys. Rev.," Vol. II, p. 450 (1913). For No. 3 I am indebted to a letter from Dr. Langmuir. Nos. 3·3, 15·1, 15·2, and 15·3 are from "Trans. Amer. Electrochem. Soc.," p. 354 (1916). Nos. 15·2 and 15·3 are stated to be preliminary measurements. No. 12, "Phys. Zeits.," Jahrg. 15, p. 525 (1914). No. 12a, "Phil. Mag.," Vol. XXIX, p. 811 (1915). Nos. 11a and 15a, "Ann. der Phys.," Vol. XLVII, p. 573 (1915). No. 14·1 "Phys. Rev.," Vol. VIII, p. 534 (1916).

Many of the values have only been worked out rather roughly as the final numbers are incapable at present of being interpreted with any great accuracy. All the values of C have not been calculated. They are in a constant proportion to the values of A_1 except for the factor $T^{\frac{3}{4}}$, T being the absolute temperature in the different experiments.

All the data in the table, except No. 10, were obtained

under conditions which, at the time when the various experiments were made, led the authors to believe that they were measuring the emissions characteristic of the elements in question. It was thus expected that the constants A and *b* would have definite values for each material. This statement is exactly true only so far as concerns *b*, which depends only on the relative values of the currents at different temperatures. In one or two of the experiments there is some latitude in the value of A, which depends on the absolute value of the currents, on account of uncertainty as to the exact area of the emitting surface, occurrence of some impact ionization, and difficulty of attaining saturation. However, in most of the experiments these uncertainties were not present, and in any event they would not be expected to affect the order of magnitude of A seriously.

A glance at the table shows that the expected result is far from having been attained. The variation of *b*, for example, for a given element is enormously greater than the variations arising from errors of measurement justify. In fact, the researches referred to at the beginning of this chapter have proved one thing with great clearness, namely, that the determination of the emission constants for the elementary substances is an experimental problem of the most extraordinary difficulty. There are two chief reasons for this.[1] In the first place the rate of emission is extremely sensitive to the minutest traces of a large number of gases. In the second place the general character of the emission from a wire subject to traces of gaseous contamination is not affected thereby. That is to say, the hot body adjusts itself to the altered circumstances, so that the emission still follows a current temperature law of the form $i = AT^{1/2} e^{-b/T}$ but with different values of the constants. There is therefore nothing in the behaviour of the phenomena itself which enables one to tell when, or if, the desired purification has been attained. Moreover, the effects produced by extremely minute amounts of gas are so considerable that it is doubtful, at the present stage of

[1] Cf. chap. IV.

development of this branch of experimentation, whether the requisite degree of purity can be attained except in the case of a small number of highly refractory elements. The most successful experiments from this point of view are those of Langmuir and K. K. Smith on tungsten, and the numbers in the table opposite this element deserve more confidence as representing values characteristic of the substance itself than any of the other numbers. The data of Langmuir and of Stoekle for molybdenum also appear deserving of confidence and are in fair mutual agreement.

The effect of gases on the emission will be considered at length in the next chapter, but it is necessary to say a few words about it in order intelligently to discuss the contents of the preceding table. The early experiments of McClelland,[1] with platinum and German silver wires, showed that the emission was unaffected when the pressure of the surrounding gas was changed from 0·004 to 0·04 mm. This rather indicated that to obtain the characteristic elementary values it was necessary only to get rid of gas to an extent sufficient to avoid complications due to secondary actions between the liberated electrons and the gas, such, for example, as impact ionization. Thus in the writer's experiments Nos. 1 and 4 no precautions to avoid gaseous contamination were taken except to keep the pressure well under 0·01 mm. by continuous pumping. Working with platinum wires H. A. Wilson[2] found that the emission had the same value at a given temperature in air, nitrogen, and water vapour at low pressures. On the other hand, the emission was enormously increased by hydrogen even when this gas was present in very small quantity. The vapours of mercury and phosphorus pentoxide were also found to increase the emission to some extent at high temperatures, although it does not seem certain that the increase caused by these vapours was not due to impact ionization. The writer[3] found later that the negative emission from platinum in an atmosphere of oxygen was

[1] " Camb. Phil. Proc.," Vol. XI, p. 296 (1901).
[2] " Phil. Trans., A.," Vol. CCII, p. 262 (1903).
[3] *Ibid.*, Vol. CCVII, p. 1 (1906).

independent of the pressure at pressures below 1 mm. when there was no impact ionization. In experiment No. 4 the gas present was that given off from the wire and the surrounding electrode under the influence of heat, and gases emitted from hot metals usually contain a considerable proportion of hydrogen. This led Wilson to think that the observed emission from platinum in general might be largely or entirely conditioned by the presence of hydrogen. He therefore sought to remove all traces of hydrogen from the wires he experimented with by boiling them in pure nitric acid and also submitting them to the action of nascent electrolytic oxygen for long periods. We shall see also that certain oxides, particularly lime, have a much greater power of emitting electrons than platinum, and if we are to observe the effects from platinum itself it is necessary to get rid of all traces of these substances. This would be accomplished by the nitric acid treatment. After a purification lasting one hour Wilson observed the values given under No. 5, whilst those under No. 6 were obtained after treatment lasting twenty-four hours. It will be seen that the effect of the treatment is to increase b very considerably. As the value of A_1 is not much affected this corresponds to a large reduction of the emission, especially at the lower temperatures. At 1500° C. the values under No. 4 gave an emission about 250,000 times as large as the values under No. 6. Notwithstanding this large reduction in the emission it does not appear that it can be got rid of entirely by removing all traces of hydrogen. This is shown very strikingly by an experiment made by the writer[1] in which a hot exhausted platinum tube was used. This tube was heated for a long time in air at atmospheric pressure and was found to give the small emission, constant at a given temperature, which characterizes a wire which has been thoroughly soaked in oxygen, as in Wilson's treatment. Hydrogen was then allowed to diffuse through the walls of the tube by admitting it to the interior. Even when relatively large amounts of hydrogen diffused out of the tube no increase in the observed negative thermionic

[1] " Phil. Trans., A.," Vol. CCVII, p. 1 (1906).

emission from the outside of the hot tube could be detected. It seems impossible to reconcile the results of this experiment with the view that the emission from platinum is entirely and fundamentally conditioned by the presence of hydrogen.

The values under No. 7 are for a platinum wire cleaned with nitric acid and heated in oxygen at a pressure of 1·47 mms. The potential difference used in making the measurements was 40 volts, and as there is some impact ionization under these conditions the value of A_1 will be a little too high thereby. Number 8 is for a clean wire but not specially oxidized. However, there is no doubt that in Deininger's experiments as a whole the apparatus was well glowed out, and they show relatively little evidence of obvious trouble from gaseous contamination. Number 9 is for a platinum wire cleaned with nitric acid and heated in an atmosphere of helium at a low pressure. No. 10 is for a clean platinum wire in hydrogen at 133 mms. and is included in order to exhibit the enormous reduction in both b and A_1 which occurs when platinum is heated in this gas. The reduction in b much more than offsets the reduction in A_1, so that the combined effect is an increase in the emission. The increase is more marked at relatively low temperatures. Number 11 is a preliminary result given by Langmuir, supposedly for very good conditions as to freedom from gaseous contamination. As the details have not yet been published it is impossible to criticize this result, but the values are widely different from those found by the other experimenters. (See, however, p. 137.)

In considering the variation of A_1 and b it is important to remember that a given variation of b means a great deal more than a variation of A_1 in the same proportion. b is deduced directly from the ratio of the currents at two known temperatures, and A_1 is then obtained from a knowledge of the thermionic current per unit area at any known temperature. On account of the exponential relation a small error in b gives rise to an enormously greater error in A_1. Thus in a particular case worked out by the writer[1] it was found that an

[1] " Phil. Trans., A.," Vol. CCI, p. 542 (1903).

error of 10 per cent in b changed A by a factor of 100, whilst an error of 33 per cent in b multiplied A by a factor of 3×10^7. Another possible source of uncertainty, depending on the form of the temperature law, arises if b is a linear function of T. Thus if $b \equiv b_0 + \beta T$ we have

$$ N = A_1 T^{\frac{1}{2}} e^{-b/T} = A_1 e^{-\beta} T^{\frac{1}{2}} e^{-b_0/T} $$

Part of the constant A_1 as given by the experiments will then arise from the temperature coefficient of b, the constant becoming in fact $A_1 e^{-\beta}$. So far as the pure metals are concerned, the theory in the last chapter indicates that the temperature variation of the quantity corresponding to b is comparatively small and, to a close approximation, calculable. The possibility of this complication has, however, to be borne in mind when we are dealing with a contaminated surface whose constitution may change with changes of temperature and other conditions.[1]

Let us now consider if it is possible to draw any conclusion as to the probable value for uncontaminated platinum from the figures given in the table. No. 10 in hydrogen at 133 mm. pressure may be at once left out of account and it is likely that No. 4 also suffers from hydrogen contamination. It is at least possible that the rather drastic oxygen treatment to which Nos. 5, 6, and 7 were subjected does more than was desired by leaving a layer of oxygen at the surface of the metal, which tends to retard the escape of the electrons. No. 11 may be left out of account pending further details as it is quite out of line with all the rest. This leaves Nos. 8 and 9 which agree with one another. There is no obvious objection to them, and it seems likely that the best guess we can make at present is that the final value of b for platinum will be somewhere near 6×10^4 and the other quantities near the values given under No. 8.[2] Schlichter,[3] however, has

---

[1] Cf. pp. 123 ff., chap. iv.

[2] Cf., however, O. W. Richardson, " Roy. Soc. Proc., A.," Vol. XCI, p. 524 (1915).

[3] Loc, cit,

recently concluded that the lower values of b given by Nos. 4 and 11a are probably nearest to the correct values for un-contaminated platinum. He has pointed out that the emission from a pure metal surface is characterized by the occurrence of saturation without accelerating potential. This criterion was satisfied by the experiment which led to the values under No. 11a whereas this test was not investigated in the other experiments. The criterion emphasized by Schlichter is an important one, but it can scarcely be regarded as an absolute guarantee that the requisite purity has been attained, since it is at least possible that a platinum surface saturated with hydrogen, for example, would satisfy this criterion and still not give the values of the emission constants characteristic of the metal. At least it seems safer to adopt some such position until the matter has been subjected to a more searching experimental test.

Turning to the values for carbon, No. 1 can safely be neglected as being affected by some serious error, probably arising from impurities in the material used. Nos. 2 and 3 agree moderately well; on the whole No. 3 should be better than No. 2. This is substantiated by No. 3·3 which is a later value.

The values for tungsten, Nos. 12 and 12a, are probably much the most reliable in the whole table. Even here, how-ever, there is a difference by a factor of almost twenty in the two values of A_1.

There is a big difference in the two sets of values for tan-talum. No. 14 is probably the more reliable. The observa-tions for molybdenum are in fairly good agreement and there is no reason to question their substantial correctness. There is also no reason to question the accuracy of Langmuir's thorium data except that no details are given in the paper.

The values under 15a are for the saturation currents from nickel as determined by Schlichter. His values of the current under zero potential difference were much lower and gave values of b nearly 50 per cent higher than those in the table. Thus the criterion referred to above was far from being satisfied and the behaviour of nickel seems to call for further examination. The iron and titanium values of Dushman are definitely stated by Langmuir to be preliminary measurements.

Horton's experiments were made with calcium sublimed on to purified platinum electrodes in helium at a low pressure. It was shown that the gas emitted by the calcium during sublimation had no measurable effect on the emission. Fredenhagen [1] has since made experiments with metallic calcium from which he concludes that the emission from this substance is caused entirely by oxidation. Even if this were a possible explanation of the large currents he obtained from calcium heated in tubes which were not completely gas-tight, it would not seem to apply to the results of Horton, who took very thorough precautions against oxidation, and also obtained much smaller currents than Fredenhagen. At any rate, the objection urged cannot be accepted without more substantial experimental support. At the same time one cannot feel very certain that the values given by Horton represent the true emission values for pure elementary calcium, since this substance is such a powerful absorbent of gases at high temperatures that it is doubtful whether the evaporation method of freeing it from gaseous contamination is satisfactory.

The writer's experiments with sodium No. 17 were made under very unsatisfactory conditions, and are not considered to have any precise quantitative significance. They were made under conditions such that there was a very considerable evolution of gas inside the apparatus, and there was no approach to saturation. Moreover, it has been shown by Haber and Just [2] that there is a very considerable emission of electrons from the alkali metals at ordinary temperatures, when they react chemically with such gases as O_2, H_2O, HCl, etc. Probably this effect is much augmented when the temperature is raised ; and Fredenhagen [3] has found that the large currents ordinarily obtained from sodium and potassium are enormously

[1] "Ber. kön. Sächs. Gesell. der Wiss. Math. Physik. Kl.," Leipzig, Vol. LXV, p. 56 (1913).

[2] "Ann. der Phys.," Vol. XXXVI, p. 308 (1911).

[3] "Verh. d. Deutsch. Physik. Ges.," Jahrg. 14, p. 384 (1912); ibid., Jahrg. 16, p. 201 (1914).

reduced by getting rid of traces of gas by continued distillation *in vacuo*. At the same time the smallest currents obtained by Fredenhagen from sodium and potassium were enormously greater than those given by the more electronegative elements like platinum and carbon at equal temperatures, and, moreover, they were not saturated. On the whole we are only justified in concluding that little is known definitely about the magnitude of the emission from these metals. No doubt the difficulty of removing traces of gas in these cases is similar to that met with in the case of calcium.

Leaving out of account the data under Nos. 1, 10, 15a, and 17, the values of b all lie between the limits 3.65×10^4 and 8×10^4. An estimate[1] of the order of magnitude of b can be got by considering the electrostatic attraction of the conductor on the escaping electron. The force on the electron is attractive, and equal to that arising from an equal charge situated at its mirror image in the surface. Its amount at distance z is thus $\epsilon^2/4z^2$. This force would be infinite at the plane $z = 0$, and if the electricity were continuously distributed in the conductor an infinite amount of work would have to be done to remove a finite quantity of electricity. Owing to the discrete distribution of the electricity, however, there is an effective lower limit to z which may be denoted by d, where d is a quantity comparable with the average distance between the electrons in the conductor.[2] The order of magnitude of the work done by an electron in escaping will thus be given by

$\frac{1}{4}\int_d^\infty \frac{\epsilon^2}{z^2}dz$. If we put $d = 5 \times 10^{-9}$ this expression gives

$b = 5 \times 10^4$ roughly, in agreement with the observed values. Such a value of d would indicate that the number of electrons in atoms is of the same order as the atomic weight, in agreement with current estimates.

The values of ϕ_0, the potential difference in volts through

[1] O. W. Richardson, " Phil. Trans., A.," Vol. CCI, p. 543 (1903).

[2] For a fuller discussion see Richardson, loc. cit. ; Schottky, " Physik. Zeits.," Vol. XV, p. 872 (1914) ; Langmuir, " Trans. Amer. Electrochem. Society," p. 377 (1916) ; J. Frenkel, " Phil. Mag.," Vol. XXXIII, p. 297 (1917).

which an electron would have to fall in order to acquire an amount of energy equal to that necessary to escape from the substance at the absolute zero of temperature, have been obtained from the relation

$$d = \frac{e\phi_0 \times 10^8}{k},$$

e being expressed in electro-magnetic units. This deduction does not require a knowledge of the absolute value of the ionic charge e. For if we multiply top and bottom by ν, the number of molecules in 1 c.c. of a perfect gas at $0°$ C. and 760 mm. pressure, we obtain, after transposing,

$$\phi_0 = d \times \frac{k}{e} \times 10^{-8} = d \times \frac{\nu k}{\nu e} \times 10^{-8}, \qquad . \quad (25)$$

where νk = R the gas constant for 1 c.c. and νe = the charge in electro-magnetic units required to liberate 0·5 c.c. of H_2 in electrolysis (0·5 c.c. since the molecule of hydrogen contains two atoms). These are both well-known physical quantities having the values

R = $3·72 \times 10^8$ erg. deg.$^{-1}$ and νe = ·4327 e.m.u.

The values of ϕ_0 are all seen to lie between 3 and 6·6 volts. So far as the order of magnitude is concerned they support the theoretical conclusion reached in Chapter II, p. 44, that the differences of ϕ_0 should be equal to the contact potentials between the different metals. The values of ϕ_0 are not, however, reliable enough adequately to test this conclusion in detail. The best support is given by the value ϕ_0 = 3·04 for calcium which, when compared with the most probable value for platinum, would make the former about 2 volts positive to the latter element. According to the experiments made by the chemical method by Wilsmore,[1] calcium is 3·42 volts electropositive to platinum, but the chemical method usually appears to give differences about 50 per cent greater than the direct contact methods. On the other hand, from the discussion in Chapter II we saw that it is not yet established with absolute certainty that there is any considerable contact

[1] " Zeits. für physik. Chemie," Vol. XXXV, p. 291 (1900); Winkelmann's " Handbuch der Physik," 2nd edition, Vol. IV, Pt. II, p. 855.

difference of potential between absolutely pure gas-free metals in a perfect vacuum.   Whether there is or is not, it is difficult to see how the theoretical relation under discussion can avoid being satisfied.   The question of the existence of contact electromotive force under ideal conditions of freedom from gaseous contamination is undoubtedly of the highest and most immediate importance from the standpoint of the theory of the emission of electrons.   Unfortunately, it is a problem which furnishes the most extraordinary experimental diffi- culties.

According to the quantum theory considered in Chapter II, the value of C (A_2 of equation (32), p. 41) should be nearly the same for all good conductors such as those included in the table on pp. 81-82.   The value of C calculated from equation (32), Chapter II, is

$$A_2 = C = 1 \cdot 5 \times 10^{10}$$

It will be observed that the good values of C given in the table are all somewhat higher than this, but do not exceed it by more than a factor of 10, approximately.   Considering the difficulty of determining C with any approach to accuracy, this agreement affords some support for the quantum theory there developed.   On the other hand, the good values of A_1 on pp. 81-82 on comparison with the corresponding constants in equation (19) of Chapter II, which is based on the classical kinetic theory, give values of n_1, the number of free electrons in 1 c.c. of the different metals, which range around 10^{21} to 10^{22} and are in agreement with the estimates from optical data.   On account of the uncertainty underlying the experimental values of A_1 and C it does not appear profitable to discuss this ques- tion further at the present time.

EMISSION OF ELECTRONS FROM COMPOUND SUBSTANCES.

The property of emitting electrons when heated is not con- fined to the list of elementary substances which constitute the conductors of the ordinary metallic type.   In fact there is no reason to doubt that the property is one which pertains to all types of matter provided that the condition of stability at the requisite high temperature is satisfied. This view is strongly supported by the facts that all known substances conduct electricity with facility at high temperatures, and all of the very large number which have been carefully examined in this respect exhibit the power of emitting electrons.

The first demonstration that this property was possessed by compound substances was given by Wehnelt.[1] In making measurements of the fall of potential at a heated platinum cathode in a discharge tube he found that the fall was greatly reduced when the hot cathode was covered with a thin layer of various oxides, notably those of calcium, strontium, and barium. The reduced cathode fall was found to be due to the increased emission of electrons from the cathode caused by the presence of the oxides. Although the oxides mentioned were much the most efficient, some reduction was also found to occur with the oxides of magnesium, zinc, cadmium, yttrium, lanthanum, thorium, and zirconium. On the other hand, the oxides of beryllium, aluminium, thallium, titanium, cerium, iron, nickel, cobalt, chromium, uranium, tin, lead, bismuth, silver, and copper showed no effect.

The emission from the oxides of the alkaline earth metals was examined in detail by Wehnelt and was found to show a close correspondence with that exhibited by the typical metallic conductors. The current E.M.F. curves were similar, showing saturation at high and low pressures, and effects due to ionization by collisions at intermediate pressures of the order of 1 mm. Careful experiments [2] have, however, since shown that the current from these cathodes never fully saturates at very low pressures: there is always a small increase with the voltage similar to that observed with freshly heated metals referred to on p. 200. The explanation of this difference is not quite certain. It may be simply that these oxide layers continue to give off gas much longer than the metals, owing

---

[1] " Sitzungsber. der physik. med. Soc. Erlangen," p. 150 (1903); " Ann. der Phys.," Vol. XIV, p. 425 (1904); " Phil. Mag.," Vol. X, p. 88 (1905).

[2] Wehnelt and Jentzch, " Verh. d. Deutsch. Physik. Ges.," Jahrg. 10, p. 605 (1908).

to more tenacious retention. On the other hand, there is no clear evidence that the effect is due to gas evolution. Wehnelt found the temperature variation of the approximately saturated current at low pressures to be governed by the same formula $i = AT^{\frac{1}{2}}e^{-b/T}$ as that from hot metals. This result has been confirmed by experiments made later by Deininger, Horton, Jentzsch, and others. The approximate values of the emission constants are given in the following table :—

Substance.	Observer.	A_1.	b.	C.	d.	ϕ_0 (volts).
BaO	Wehnelt[1]	7×10^{26}	$4 \cdot 5 \times 10^4$	—	—	$3 \cdot 65$
CaO	Wehnelt[1]	$4 \cdot 5 \times 10^{26}$	$4 \cdot 3 \times 10^4$	—	—	$3 \cdot 48$
CaO	Deininger[2]	$1 \cdot 1 \times 10^{26}$	$4 \cdot 3 \times 10^4$	$2 \cdot 65 \times 10^{11}$	$4 \cdot 05 \times 10^4$	$3 \cdot 48$
CaO	Horton[3]	4×10^{30}	$4 \cdot 8 \times 10^4$	—	—	$3 \cdot 9$
CaO	Jentzsch[4]	$4 \cdot 3 \times 10^{26}$	$4 \cdot 03 \times 10^4$	—	—	$3 \cdot 36$

The values of the constants for CaO agree quite well except Horton's. In the experiments of Wehnelt, Deininger, and Jentzsch the layer of oxide was deposited by evaporating a solution of calcium nitrate and then heating the calcium nitrate until it turned into the oxide. In Horton's experiments, which had a different objective from that of the others, the lime was prepared by the action of oxygen gas on a hot surface of metallic calcium. The measurements were made under a potential difference of 40 volts in an atmosphere of helium at 3·24 mm. pressure. No doubt these circumstances would affect the measurements in various ways. Horton found that at all temperatures between 700° C. and 1400° C. the emission from lime was much greater than that from calcium. A comparison with the data in the table on p. 82 shows that the difference between the effect from calcium and that from lime prepared from calcium nitrate is not so great as that obtained when calcium is compared with lime prepared by oxidation.

A serious difficulty in experimenting with metals coated with oxides has been caused by the splitting off or disintegration of the oxide. This difficulty is overcome in the coated

[1] "Ann. der Phys.," Vol. XIV, p. 425 (1904).
[2] *Ibid.*, Vol. XXV, p. 285 (1908).
[3] "Phil. Trans., A.," Vol. CCVII, p. 149 (1907).
[4] "Ann. der Phys.," Vol. XXVII, p. 129 (1908).

filaments prepared by the Western Electric Co. of America [1] where the oxides of calcium, strontium, or barium or a mixture of them are laid on in successive thin coats in the form of hydroxide or carbonate mixed with some carrier such as resin or paraffin which is afterwards burnt away in the air.

The experiments of Deininger [2] are particularly instructive and very consistent. He measured the emission at various temperatures from filaments of platinum, carbon, tantalum, and nickel both in the ordinary state and then when coated with lime from calcium nitrate, and found that the emission from the lime-covered wires at a given temperature was the same in all cases.

This shows that in these experiments we are dealing with a definite property of the oxides which is quite independent of the underlying metal. The excellence of the agreement may be judged from Fig. 9, p. 63, where a few of the observations with lime on platinum, tantalum, and nickel respectively are plotted. The values have been selected so as to exhibit the worst as well as the best agreement.

The emission from Nernst filaments has been carefully examined by Owen [3] and Horton. [4] It is smaller than that from the alkaline earths, otherwise there are no special features. From the data found by Owen the approximate values of the constants are $A_1 = 7 \times 10^{23}$ and $b = 4 \cdot 6 \times 10^4$.

A systematic examination of a large number of metallic oxides has been made by Jentzsch. [5] The oxides were deposited on platinum wires by the decomposition of appropriate solutions. By taking very great care in the purification of the platinum the emission from the metal was reduced to a low value and the number of oxides which were found capable of a greater emission, at relatively low temperatures, was greatly extended. The emission from practically all of these substances increases with temperature more slowly than that from

[1] H. D. Arnold, " Phys. Rev.," Vol. XVI, p. 73 (1920).
[2] Loc. cit.
[3] " Phil. Mag.," Vol. VIII, p. 330 (1904).
[4] " Phil. Trans., A.," Vol. CCXIV, p. 277 (1914).
[5] " Ann. der Phys.," Vol. XXVII, p. 129 (1908).

platinum itself; so that even if they have a greater power of emission at low temperatures, at sufficiently high temperatures the platinum will catch up to and overtake them. In the case of zinc and magnesium oxides Jentzsch found that this happened at about 1600° C. Of the oxides tested the only ones which were found to give rise to no effect were those of thallium and lead which probably volatilized before sufficiently high temperatures could be attained. In the case of lithium oxide some emission was observed between 700° C. and 800° C., but this disappeared on raising the temperature further, also probably on account of volatilization. The values of the constants deduced by Jentzsch from his measurements with different oxides are collected in the following table :—

		Values of		
Oxide of	A_1.	n_1.	b.	ϕ (volts).
Ba	$2 \cdot 94 \times 10^{26}$	$2 \cdot 0 \times 10^{22}$	$4 \cdot 16 \times 10^4$	$3 \cdot 58$
Sr	$3 \cdot 16 \times 10^{26}$	$2 \cdot 1 \times 10^{22}$	$4 \cdot 49$,,	$3 \cdot 87$
Ca	$2 \cdot 68 \times 10^{26}$	$1 \cdot 8 \times 10^{22}$	$4 \cdot 03$,,	$3 \cdot 48$
Mg	$2 \cdot 11 \times 10^{19}$	$1 \cdot 4 \times 10^{14}$	$3 \cdot 95$,,	$3 \cdot 40$
Be	$6 \cdot 45 \times 10^{18}$	$4 \cdot 3 \times 10^{13}$	$2 \cdot 39$,,	$2 \cdot 06$
Y	$1 \cdot 17 \times 10^{23}$	$7 \cdot 8 \times 10^{17}$	$3 \cdot 63$,,	$3 \cdot 13$
La	$4 \cdot 3 \times 10^{21}$	$2 \cdot 9 \times 10^{16}$	$3 \cdot 79$,,	$3 \cdot 26$
Al	$4 \cdot 0 \times 10^{19}$	$2 \cdot 7 \times 10^{14}$	$3 \cdot 73$,,	$3 \cdot 21$
Zr	$4 \cdot 1 \times 10^{22}$	$2 \cdot 7 \times 10^{17}$	$3 \cdot 66$,,	$3 \cdot 15$
Th	$2 \cdot 19 \times 10^{20}$	$1 \cdot 5 \times 10^{15}$	$3 \cdot 56$,,	$3 \cdot 06$
Ce	$1 \cdot 22 \times 10^{22}$	$8 \cdot 2 \times 10^{16}$	$3 \cdot 71$,,	$3 \cdot 20$
Zn	$1 \cdot 92 \times 10^{18}$	$1 \cdot 3 \times 10^{13}$	$3 \cdot 51$,,	$3 \cdot 02$
Fe	$2 \cdot 23 \times 10^{22}$	$1 \cdot 5 \times 10^{17}$	$4 \cdot 69$,,	$4 \cdot 04$
Ni	$1 \cdot 74 \times 10^{23}$	$1 \cdot 2 \times 10^{18}$	$5 \cdot 12$,,	$4 \cdot 41$
Co	$3 \cdot 32 \times 10^{22}$	$2 \cdot 2 \times 10^{17}$	$4 \cdot 97$,,	$4 \cdot 28$
Cd	$2 \cdot 33 \times 10^{18}$	$1 \cdot 6 \times 10^{13}$	$3 \cdot 02$,,	$2 \cdot 60$
Cu	$2 \cdot 19 \times 10^{16}$	$1 \cdot 5 \times 10^{11}$	$2 \cdot 25$,,	$1 \cdot 94$

The values of ϕ have been calculated directly from the relation $\phi = b\dfrac{R}{e} \times 10^{-8}$ and have not been reduced to the absolute zero. They are therefore a little larger than the corresponding values of ϕ_0. The differences are, however, small. The values are clearly much smaller than those for the refractory metallic conductors, corresponding to the smaller rate of increase of emission with rising temperature. The values of n_1 as calculated from equation (19), Chap. II, have also been included. These numbers cannot, however, be regarded as the number of free electrons present in unit volume of the oxides. This is evident from the following considerations.

The low electrical conductivity of the incandescent oxides compared with that of the metals indicates that the classical kinetic theory will probably apply both to the internal and to the external free electrons in these cases. Admitting this, let ν_0 be the concentration of the external, and ν_1 of the internal, free electrons, then by a well-known theorem [1]

$$\nu_0 = \nu_1 e^{-w/RT} \qquad . \qquad . \qquad . \qquad . \qquad (26)$$

where w is the work done by an internal free electron in escaping. Now we know from thermodynamics that ν_0 is of the form

$$\text{const.} \times e^{\int \frac{L}{RT^2} dT}$$

where L is latent heat of evaporation, a quantity whose temperature variation may to a first approximation be disregarded. Thus ν_0 is very close to the form const. $\times e^{-\text{const.}/T}$, and since the variation of w with T may be disregarded (a linear variation makes no difference), and since also ν_1 does vary rapidly with T, as is shown by experiments on the electrical conductivity of heated oxides, ν_1 must be of the same form, and given by, let us say, $\nu_1 = A' e^{-w'/RT}$ where A' and w' are constants. Hence from (26)

$$\nu_0 = A'e^{-(w+w')/RT} \qquad . \qquad . \qquad . \qquad (27)$$

Thus the process of taking out the exponential temperature factor, which is what the calculation of n_1 really amounts to, leaves neither ν_1 nor n_1 but A' an arbitrary constant. This argument is only an approximate one and may perhaps appear involved, but there is no doubt that the conclusion is sound. The enormous temperature variation of the conductivity of metallic oxides alone is sufficient to show that n_1 is not constant as the numbers in the table would indicate. It is interesting to observe that the values of A_1 given by the oxides generally are lower than those given by the metals, the alkaline earths alone being in the same class with the metals in this respect.

[1] Cf. O. W. Richardson, " Phil. Mag.," Vol. XXIII, p. 608 (1912).

All the oxides for which the constants are tabulated behave quite regularly, but the emission from manganese oxide was found by Jentzsch to be peculiar. As the temperature was raised there was found to be a sudden increase in the emission at a certain stage and this increased value of the emission was found to persist when the temperature was subsequently reduced. A second sudden increase to a still higher value of the emission was detected when the temperature was raised to a value higher than any previously employed. These effects are attributed by Jentzsch to the formation of the various oxides of manganese.

The emission from Wehnelt cathodes—hot metal cathodes coated with lime or baryta—has attracted a good deal of attention, partly owing to their practical application as a convenient source of powerful electron currents. Fredenhagen[1] has described a number of experiments which led him to the conclusion that the emission of electrons from these cathodes is a secondary effect, arising from the recombination of the earth metal with the oxygen liberated by electrolysis during the passage of the current through the oxide. This view acquires a certain amount of plausibility owing to the results of a research of Horton's[2] on the electrical conductivity of heated oxides, in which he concludes that such conductivity, although mainly electronic (i.e. of the same type as that of metals), is to some extent accompanied by electrolysis. One result of Fredenhagen's view is that the electronic emission from lime at a given temperature should be larger when the lime is heated electrically than when methods of heating which do not involve the use of an electric current are employed. This test is difficult to perform satisfactorily and the earlier experiments seemed to indicate such a difference. In later experiments made under better conditions Fredenhagen[3] was able to heat a lime-coated platinum strip by means of (1) an electric current through the strip and (2) a beam of heat radiation focussed

[1] "Ber. d. Sächs. Ges. d. Wiss. Math. Physik. Kl.," Vol. LXV, p. 42 (1913); "Phys. Zeits.," Jahrg. 15, p. 21 (1914).

[2] "Phil. Mag.," Vol. XI, p. 505 (1906).

[3] "Ber. d. Sächs. Ges. d. Wiss. Math. Physik. Kl.," Vol. LXV, p. 55 (1913).

on the surface of the lime. This experiment showed no difference in the emissions at a given temperature, except what might be due to unavoidable experimental error. The fact that the mode of heating makes no difference in the emission at a given temperature has been confirmed by Horton[1] by experiments with the Nernst glower. The other grounds on which Fredenhagen rests his thesis may briefly be summarized as follows :—

1. The oxide gradually disappears as the cathode is worked. It probably disappears slowly when it is heated without emission occurring, but the disappearance is undoubtedly much faster when emission is occurring, i.e. when the oxide is negatively charged.

2. Gas is given off when the cathode is in action.

3. The underlying platinum shows corrosion, which Fredenhagen attributes to the formation of an alloy between the platinum and the calcium liberated by electrolysis.

4. The emission from calcium in a good vacuum is much smaller than when it is measured in a tube into which air slowly leaks.

As regards (1), Wehnelt and Liebreich[2] have brought forward very strong evidence that the loss of oxide is due to a combination of simple evaporation and sputtering due to bombardment of the oxide by positive ions arising from impact ionization. The second cause is operative only when the oxide is negatively charged, thus accounting for the increased rate of loss under these conditions.

The same authors have also considered (2) and from their experiments conclude that the gas is given off mainly in the early stages of heating. A spectroscopic examination showed the presence only of hydrogen, probably arising from the platinum and from water occluded in the lime.[3] Gehrts has observed spectroscopic evidence of the presence of calcium and oxygen in the glow from lime cathodes, but only under conditions such as would lead to extensive evaporation or

---

[1] "Phil. Trans., A.," Vol. CCXIV, p. 277 (1914).
[2] "Phys. Zeits.," Jahrg. 15, p. 557 (1914).
[3] "Verh. der Deutsch. Physik. Ges.," Jahrg. 15, p. 1047 (1913).

sputtering of the lime owing to intense cathodic bombardment by positive ions.

Tests made by Wehnelt and Liebreich show that the platinum corrodes to the same extent whether it is covered with lime or not.

The cogency of (4) appears to be disposed of by the results of the experiments of Horton, who showed that the emission from lime was much greater at a given temperature than that from metallic calcium. Horton[1] has also shown by direct experiment that there is no measurable electronic emission when calcium is oxidized at 500° C. to 600° C., although there is a very marked emission when the oxide formed is heated to 700° C. to 800° C. subsequently. This experiment would seem to prove that the act of oxidation is not an important factor as compared with the effect of change of temperature.

Taking all the evidence together it seems to the writer that the view which attributes the emission from metallic oxides to the escape, owing to increased kinetic energy, of those electrons which give rise to the electrical conductivity of such materials has much more to be said for it than any other so far put forward. This position is strengthened by the recent experiments of Germershausen,[2] who has shown that the removal of the last traces of gas from a Wehnelt cathode and its surroundings increases the emission from it. Under these conditions the discharge from the lime becomes very steady and shows temperature and voltage characteristics similar to those exhibited by tungsten filaments under the best vacuum conditions. (See pp. 68 and 132.)

In recent years oxide coated cathodes have been very thoroughly studied in the Research Laboratory of the American Telephone and Western Electric Companies.[3] One result of these researches has been to establish thoroughly the view that the emission from such cathodes is an intrinsic property of the oxide similar to the emission from pure metals and is not a secondary phenomenon due to chemical action,

[1] " Phil. Trans., A.," Vol. CCXIV, p. 292 (1914).
[2] " Phys. Zeits.," Jahrg. 16, p. 104 (1915).
[3] H. D. Arnold, " Phys. Rev.," Vol. XVI, p. 70 (1920).

bombardment of the cathode by positive ions, or the like. Tests have shown that the oxycathodes maintain their emitting power unimpaired *in vacuo* in which the measured pressures were as low as 10^{-9} to 10^{-10} mm. Filaments have been maintained in operation for such long periods that the mass of the electrons emitted by them exceeded fifteen times the mass of the oxide coating. It appears that the value of b for filaments coated with a mixture of the oxides of barium, strontium, and calcium is much less than that for a pure lime filament and their emitting power at a given temperature is correspondingly greater. The value of b deduced from the temperature variation of the emission is very close to the value deduced from the cooling effect[1] (see Chap. V, p. 187). Direct tests by Davisson and Germer[2] show that any electron emission which may be due to positive ion bombardment must in general under the conditions in which the tubes are operated be under one ten-thousandth of the total emission and such effects can therefore play no important part in the phenomena. Very interesting experiments[3] have been made with filaments coated with oxide evaporated from a second oxide coated filament. It appears that exceedingly small amounts of the oxide (enough to cover the underlying metal to about 30 per cent of its area with a layer of oxide one molecule deep) are sufficient to increase the emission from the value appropriate to the pure metal to that characteristic of the oxide. This corresponds to an increase in the current by a factor of about 10^9 under the conditions of the experiments. The variation of the emission with the amount of oxide deposited has been investigated. It is found that for very minute deposits the emission practically remains at the value for the pure metal, as the deposit increases the emission begins to rise rapidly in two steps to a maximum value after which it falls with similar rapidity to the value characteristic of a thick oxide layer. Up to beyond the maximum the emission can be represented as

[1] Wilson, " Phys. Rev.," Vol. X, p. 79 (1917).

[2] Davisson and Germer, " Phys. Rev.," Vol. XV, p. 330 (1920).

[3] Arnold, loc. cit., and Davisson and Pidgeon, " Phys. Rev.," Vol. XV, p. 553 (1920).

the sum of two terms of the form $Cx^n e^{x+a}$ when C, n, and a are constants and x is the amount deposited. The drop in the emission over the falling part beyond the maximum appears to be due to a drop in the value of A from that for the pure metal to the smaller value appropriate to the oxide, the other constant b in the emission formula remaining unaltered. Apparently the thinnest deposit has already established the value of b appropriate to the oxide.

Horton[1] and Martyn[2] found that the emission from the Wehnelt cathode, like that from a hot platinum wire, was greatly increased in an atmosphere of hydrogen.[3] Martyn, whose experiments were made in air and hydrogen at atmospheric pressure, found that when the currents were approximately saturated the results could be expressed in a very simple manner. If at a given temperature x is the thermionic current from a clean platinum wire in air, ax that from a lime coated platinum wire in air, and bx that from a clean platinum wire in hydrogen, then the thermionic current from a lime coated platinum wire in hydrogen at the same temperature is abx. a and b were both found to be very nearly equal to 10^5 at 1600° C., and varied to some extent with the temperature.

This result, which has only been established approximately, appears to require, to a corresponding degree of approximation, that the contact potential ·difference between lime in an atmosphere of ·hydrogen and lime in air should be equal to that between platinum in hydrogen and platinum in air, when the temperatures are the same in both cases. This follows from a consideration of the equilibrium of the electrons in an enclosure containing bodies of lime and platinum in suitable electrical connexion. The enclosure is imagined to be divided into two separate parts by a diaphragm permeable to electrons

---

[1] " Phil. Trans., A.," Vol. CCVII, p. 149 (1907).

[2] " Phil. Mag.," Vol. XIV, p. 306 (1907).

[3] Horton (" Roy. Soc. Proc., A.," Vol. XCI, p. 322 (1915)) has recently concluded that this increase occurs to any appreciable extent only when the pressure of the hydrogen is considerable and that it is perhaps to be attributed to an interaction between the hydrogen and the platinum. At pressures comparable with 0·01 mm. he found little difference in the emissions from lime or the Nernst glower in hydrogen, air, oxygen, or nitrogen.

but not to gases, one part containing hydrogen at a definite pressure and the other containing air or exhausted. The partial pressure of the electrons is the same on both sides of the diaphragm whose presence cannot affect the conditions which determine their equilibrium. Let n_1 and V_1 respectively denote the equilibrium concentration and the potential of the electrons just outside the platinum in air, n_2 and V_2 the corresponding quantities for lime in air, n_1' and V_1' for the platinum in hydrogen, and n_2' and V_2' for the lime in hydrogen. Then, as in Chapter II, p. 42, the condition of equilibrium of the electrons in the enclosure requires that

$$n_1/n_2 = e^{-\epsilon(V_1-V_2)/kT} \qquad . \qquad . \qquad . \quad (28)$$
$$n_2'/n_2 = e^{-\epsilon(V_2'-V_2)/kT} \qquad . \qquad . \qquad . \quad (29)$$
$$n_1'/n_2 = e^{-\epsilon(V_1'-V_2)/kT} \qquad . \qquad . \qquad . \quad (30)$$

But if we neglect electron reflexion, since the saturation currents i are proportional to the corresponding values of n at constant temperature,

$$n_1/n_2 = i_1/i_2 = a, \qquad . \qquad . \qquad . \quad (31)$$

and

$$n_2'/n_2 = i_2'/i_2 = b \qquad . \qquad . \qquad . \quad (32)$$

If Martyn's result is to hold

$$i_1'/i_2 = n_1'/n_2 = ab ; \qquad . \qquad . \qquad . \quad (33)$$

so that, from (28), (29), and (30),

$$V_1' - V_1 = V_2' - V_2, \qquad . \qquad . \qquad . \quad (34)$$

which is the condition referred to. Since from (34)

$$V_1' - V_2' = V_1 - V_2, \qquad . \qquad . \qquad . \quad (35)$$

this condition is also embodied in the statement that, to the degree of accuracy in question, the contact potential difference between platinum and lime should be the same in air as in hydrogen. The contact differences mentioned are those which would obtain at the temperatures of the experiments, not at ordinary temperatures.

As to the nature of the process by which the hydrogen affects the emission, and, according to the foregoing theory, the contact differences of potential also, this may be tentatively attributed to an effect of positive hydrogen ions dissolved in the solids. The writer[1] has pointed out there is considerable evidence in favour of the view that some of the hydrogen which dissolves in platinum is not merely dissociated into atoms but exists in solution in the form of positive ions. If this is admitted it follows from the laws of chemical equilibrium that there will also be a certain concentration of hydrogen ions in the external hydrogen, and the same may be expected also of the hydrogen which is entangled or absorbed in the layer of lime. If p_1, p_2, etc., are the partial pressures of the hydrogen ions in equilibrium in the various phases of the system, then, as Sir J. J. Thomson[2] has pointed out, the difference of concentration in any two phases will give rise to a difference of potential across the interface equal to $\dfrac{k\mathrm{T}}{e} \log p_1/p_2$, etc., just as a similar term in the electromotive force arises in the theory of concentration cells. Until further information is available it seems most satisfactory to attribute the changes in contact potential difference and electron emission brought about by gases to an effect of this character. The matter will be discussed again in the next chapter, pp. 123 ff.

The discharge from hot lime cathodes is affected by other gases as well as by hydrogen. This is shown, for example, by data recently published by Fredenhagen.[3] The phenomena, except at very low pressures, are complicated by effects arising from a number of different causes, and it is impossible to disentangle the details in the data at present available. Some very interesting features of the discharge from a Wehnelt cathode in gases at low pressures have been studied by Sir J. J. Thomson.[4] In air at about 0·2 mm. pressure, for example, there is a very rapid increase in the current with increasing potential at relatively low potentials. At a certain stage (at 37 volts potential difference in one of the tubes used) a faint glow appears at the anode. This glow gradually be-

---

[1] "Phil. Trans., A.," Vol. CCVII, p. 1 (1906).

[2] "Conduction of Electricity through Gases," 2nd edition, p. 204 (Cambridge, 1906).

[3] "Phys. Zeits.," Jahrg. 15, p. 19 (1914). Cf., however, Horton, "Roy. Soc. Proc.," loc. cit.

[4] "Conduction of Electricity through Gases," 2nd edition, p. 478.

comes more extensive and brighter as the potential difference and current are increased. With a constant potential difference at this stage the currents increase with increasing distance between the electrodes, showing that part of the current arises from impact ionization, or at least from some secondary phenomenon, in the gas. At a slightly higher potential (at 53 volts in a particular instance) the luminosity, hitherto confined to the neighbourhood of the anode, suddenly extends throughout the tube. This discharge, which appears very sharply at a quite definite potential, is accompanied by an enormous increase in the current. At this stage the discharge is an ordinary luminous discharge, and the small potential difference required to maintain it corresponds to the reduction in the cathode fall originally discovered by Wehnelt.

These large currents with relatively small potential differences are attributed by Thomson to ionization by repeated impact. He supposes that if a molecule is struck several times in succession by an ion the energy is stored up in the molecule: so that ionization by collision will occur with smaller potential differences than when single impacts alone are operative. In other words, with a given potential difference on the tube impact ionization will be much more frequent with large than with small currents.

Experiments on this subject have been made more recently by Child,[1] who considers Thomson's explanation inadequate on the grounds that the phenomena occur under conditions such that the frequency of occurrence of repeated impacts is prohibitively small, and also because he concludes from his experiments that the effects are determined rather by the nature and temperature of the cathode than by the magnitude of the primary thermionic current. Thus with hot platinum cathodes much larger primary currents are required to produce the effects than with hot lime cathodes. He attributes the large currents to the emission of electrons from the cathode under the influence of the bombardment by positive ions liberated by single impact ionization in the gas.

This subject is an important one from the standpoint of

[1] " Phys. Rev.," Vol. XXXII, p. 492 (1911).

the mechanism of the ordinary luminous discharge and of the arc, and in view of the disagreement referred to there seems to be room for further experiment.

One of the great advantages of the hot lime cathode is that it may be used to furnish an intense source of electrons of very small linear dimensions. Thus if a minute speck of lime is deposited on a piece of platinum wire or foil which is heated, the emission per unit area from the lime is so enormously greater than that from the metal that, even when the disparity in area of the surfaces is taken into account, the total emission from the metal may be neglected in comparison. Very narrow streams of electrons moving with a definite velocity may be produced in this way. The application of this possibility to the measurement of e/m for the electrons, as developed by Wehnelt and perfected by Bestelmeyer, has already been referred to (p. 11). Another advantage of the cathode is that it is chemically stable in presence of most of the commoner gases. For work in high vacua the writer has found that filaments of tungsten or osmium are preferable to hot lime as a source of intense streams of electrons, owing to their greater permanency and freedom from emission of gas. They can, however, only be used under very good vacuum conditions or with inert gases, as a trace of oxygen is sufficient to eat them up almost instantaneously.

J. Lilienfeld[1] has investigated the potential gradient required to drive electron currents of considerable magnitude through long tubes under good vacuum conditions. He finds that over a considerable range the potential gradient is nearly proportional to the square root of the current and is constant along the length of the tube. If the last-named result were strictly true it would follow from Poisson's equation that the volume density of the electrification must be zero during the passage of the current. To explain his results Lilienfeld adopts the rather heroic hypothesis that the negative ions are compensated for by the presence of positive ions formed by the dissociation of the vaccum. It appears, however, that under the conditions of these experiments, even if there are

[1] "Ann. der Physik," Vol. XXXII, p. 675 (1910); Vol. XLIII, p. 24 (1914).

no positive ions, the volume density of the negative electrification would not be large, and it is doubtful if the measurements of the potential gradient are accurate enough to detect the expected variation along the length of the discharge. The difficulty of determining the local potential by means of sounding wires in a unipolar discharge is well known. Minute traces of gas would also greatly reduce the negative volume density, and it may not have been possible completely to remove them even with the very elaborate precautions in this respect which were taken by the author. At any rate it is clear that there are a number of possible explanations of a rather ordinary character which have still to be disproved. In fact a careful examination of these papers does not reveal any conclusive evidence that the potential gradient was constant along the path of the discharge under high vacua conditions.

THE EMISSION OF ELECTRONS FROM VARIOUS COMPOUNDS.

The power of emitting electrons when heated is not confined to the oxides and elementary substances. It is probably a common property of all forms of matter which are stable enough to continue in existence at sufficiently high temperatures. The writer [1] found that the following salts emitted electrons at comparatively low temperatures, viz. : the iodides of calcium, strontium, barium, and cadmium, calcium fluoride, calcium bromide, manganous chloride and ferric chloride. The emission from these salts possess important features which are not exhibited by the substances hitherto considered.

The iodides of the alkaline earth metals are remarkable for the large magnitude of the emission at relatively low temperatures. A specimen of barium iodide, heated on a platinum strip of which it covered a few square millimetres, was found to give a current of two milliamperes at a temperature so low that the strip was invisible in an ordinarily lighted room. With all the salts mentioned the emission consists in general of a mixture of electrons and negative ions of atomic

[1] " Phil. Mag.," Vol. XXVI, p. 458 (1913).

magnitude, the proportion between the two varying with the temperature and other conditions.

The behaviour of calcium iodide appears to be typical of that of the iodides of the other alkaline earth metals. At low temperatures, when first heated, the emission consists entirely of heavy ions. With freshly heated specimens of the salt no electrons could be detected at temperatures between 325° C. and 523° C. The value of ϵ/m for these heavy ions was measured. The mean of 4 determinations gave $\epsilon/m = 80.7$. This corresponds to an electric atomic weight of 120. As the chemical equivalent weight of iodine is 127, the heavy ions are evidently atoms of iodine in combination with a negative electron. If the fresh salt is heated continuously at a constant low temperature the emission increases rapidly to a maximum in about fifteen minutes and then slowly decreases. This increase to a maximum and subsequent decrease is found to occur also at the higher temperatures at which electrons are present. A similar phenomenon has been found to characterize the emission of positive ions from salts heated on strips of metal (see Chap. VIII, p. 260).

At 534° C. there was no certain evidence of the presence of electrons on first heating, but they began to be detectable immediately after passing the maximum. After two hours the current had decayed very considerably and the electrons carried about 30 per cent of the total current. At 654° C. the electrons were detectable at the outset and reached their maximum before the heavy ions. After fifteen minutes the electrons carried about 85 per cent of the current and after two hours about 60 per cent. In general, however, the proportion of the current carried by electrons increases both with duration of heating and with rising temperature. As in other cases of thermionic emission the emission of both ions and electrons tends to increase rapidly with rising temperature, other things being equal. It was noticed that the currents were smaller after the cold salt had been left in a vacuum and greater after similar exposure to air at atmospheric pressure. Measurements with strontium and barium iodides indicated that the heavy negative ions from these bodies also were iodine atoms combined with an electron.

Experiments with calcium bromide showed that the whole current from this substance at low temperatures was carried by heavy ions initially. The value of ϵ/m indicated that these ions were atoms of bromine combined with an electron.

In the case of ferric chloride the negative emission did not last long enough for measurements to be made with it. The emission from cadmium iodide also was of a temporary character. It was possible to show that both heavy ions and electrons were present in the emission but not to measure the value of ϵ/m for the former. In the case of calcium fluoride the bulk of the current was carried by electrons. When there are too many electrons present the method used for measuring ϵ/m for the heavy ions gives unsatisfactory results, and in the case of calcium fluoride all we can say is that the heavy ions were of molecular or atomic dimensions. Manganous chloride gave off both electrons and heavy ions. The electric atomic weight found for the latter had values ranging from 59 to 88. These irregular results were probably affected by the presence of too many electrons. They show, however, that the heavy ions are of molecular or atomic magnitude.

An interesting question arises as to whether the negative 'atomions' are emitted from the salts as such or are formed by the combination of electrons with dissociated atoms of the haloids subsequent to emission. This question cannot be answered with certainty; but the fact that over a considerable range of temperature no electrons can be detected when the salts are first heated, although considerable currents may be carried by negative atomions, rather favours the view that the latter are emitted as such. On the other hand one would expect negative ions sometimes to be formed by the union of electrons with uncharged atoms and molecules. Sir J. J. Thomson[1] has brought forward evidence of such processes in vacuum tube discharges at low pressures, and some of the effects of gases on the negative thermionic currents from

[1] " Roy. Soc. Proc., A.," Vol. LXXXIX, p. 1 (1913).

carbon which have been observed by Pring[1] may be accounted for[2] in this way. The negative discharge from hot platinum and tungsten in a good vacuum is purely electronic or, at any rate, the percentage of heavy ions present is too small to be detected (cf. p. 9).

THE COMPLETE PHOTOELECTRIC EMISSION.

It is well known that when light of sufficiently short wave-length is allowed to fall on metals an emission of electrons takes place under its influence. Since all substances emit light when they are raised to a high temperature an emission of electrons from hot bodies owing to the action of light will occur, even when they are not illuminated from an external source. The emission of electrons which arises in this way may be termed either the auto-photoelectric emission, to indicate that it is caused by the light radiation supplied by the hot body itself, or the complete photoelectric emission, to indicate that it is excited by the complete (black body) radiation with which the material is in equilibrium at the temperature under consideration. By applying thermodynamical principles to the equilibrium of the electrons liberated in this way by a body maintained in a vacuous enclosure at constant temperature, the writer[3] has shown that the number n of liberated electrons present in unit volume in the state of equilibrium depends upon the temperature T according to the equation

$$n = Ae^{\int^{T} \frac{\phi}{kT^2} dT}, \qquad . \qquad . \qquad . \qquad (36)$$

where A is a constant independent of T but characteristic of the substance, ϕ is the change of energy accompanying the liberation of one electron, and k is Boltzmann's constant. This equation is identical with equation (7) of Chap. II, but the constant A and the energy ϕ may not have the same values in the two cases even for the same substance. The fact that photoelectric emission under a given illumination

[1] "Roy. Soc. Proc., A.," Vol. LXXXIX, p. 344 (1913).
[2] Cf. O. W. Richardson, Vol. XC, p. 177 (1914).
[3] "Phil. Mag.," Vol. XXIII, p. 618 (1912).

is practically independent of the temperature of the illuminated substance shows that the photoelectric ϕ varies very little with T. It follows from this, together with (36), by an argument quite similar to that used in the thermionic case, that the auto-photoelectric saturation *current* is governed by the formula

$$i = AT^\lambda e^{-b/T}, \quad . \quad . \quad . \quad . \quad (37)$$

where A and b are constants and the index λ does not differ much from unity. Thus, so far as the variation with temperature is concerned, there is no clear difference between the thermionic and auto-photoelectric currents, and it is possible that the observed thermionic currents may all be attributable to photoelectric activity. This point has also been brought out by W. Wilson[1] who has calculated the emission on the assumption that all the radiant energy absorbed by hot bodies is converted, by quanta, into the kinetic energy of electrons ejected from the atoms. In this way an expression for the thermionic current is obtained which is practically the same as (37) and agrees satisfactorily with the experimental results, so far as the variation of thermionic current with temperature is concerned.

Wilson[2] has recently confirmed these conclusions by experiment. He has shown that the saturation photoelectric current from the liquid alloy of sodium and potassium when excited by approximately black body radiation is expressed by a formula of type (37), where T represents the temperature of the source of radiation. The constant b is also found to be in reasonably good accordance with the values given by purely thermionic data for the alkali metals.

It is of the utmost importance to settle whether thermionic emission of electrons is due to photoelectric action or not. As the auto-photoelectric current and the thermionic current both vary in the same way with the only controllable variable, the temperature, the only method available for deciding this question is to find whether the values of the auto-photoelectric current at a given temperature, as calculated from photoelectric

[1] " Ann. der Physik," Vol. XLII, p. 1154 (1913).
[2] " Roy. Soc. Proc., A.," Vol. XCIII, p. 359 (1917).

data, agree with the observed thermionic current at that temperature. Such indications as are available point to the conclusion that the auto-photoelectric currents are smaller than the observed thermionic currents, although it is impossible to give a completely decisive answer at present. The question is beset with extraordinary difficulties, arising from various causes. The most important of these are the difficulty of determining the correct values of either the photoelectric or the thermionic currents under given conditions, ignorance of the precise mechanism of photoelectric emission and resulting doubtfulness as to the validity of any hypotheses which may be taken as the basis of calculation, very serious mathematical difficulties attending the rigorous treatment of the theoretical problems involved, and the difficulty of obtaining some of the necessary data in an exact form.

Subject, more or less, to these reservations an idea of the magnitude of the auto-photoelectric emission from, let us say, platinum may be obtained as follows :—

Data are now available[1] which give the number of electrons emitted from platinum when unit light energy of the different effective frequencies falls on it at normal incidence. The magnitude of the auto-photoelectric emission will not, however, be obtained if we simply multiply this number by the corresponding intensity in the "black-body" spectrum and integrate the product over the whole range of frequency, on account of the different optical conditions in the two cases. In the photoelectric experiments, in which a beam of light is incident normally, the intensity of the exciting illumination is greatest at the surface and falls off exponentially as the depth of penetration increases. In the natural emission, on the other hand, the electromagnetic radiation is isotropic and its intensity is the same at all depths. This difference between the two cases can be allowed for if we have a knowledge of the coefficients of absorption of the electromagnetic radiations of various wave-lengths and of the electrons which they cause to be emitted.

[1] Richardson and Rogers, " Phil. Mag.," Vol. XXIX, p. 618 (1915).

In this connexion it is desirable to emphasize the fact that the assumption of an exponential law of absorption, for the very slowly moving electrons with which we are now concerned, can only be regarded as the very roughest kind of approximation. The problem involved here is in reality a very complex one; the stream of electrons which travel in a given direction suffer loss of their number both through true absorption and through scattering, and also lose energy as well. Little is known definitely either as to the relative importance or as to the precise effect of these different actions. In addition, the electrons which escape lose most of their energy in passing through the surface; although this fact need not prevent their rate of loss being approximately exponential when they are travelling in the interior. In any event, the exponential law of absorption is the only assumption with which it is possible to arrive at any result in the present state of the subject.

Considering the case of a beam of light of definite frequency incident normally, let I be the energy crossing unit area just within the surface of the metal in unit time. Let a be the coefficient of absorption of the liberated electrons when in the metal, β the coefficient of absorption of the light, and N the number of electrons ejected from the atoms of the metal when unit energy is absorbed by them from the light. Only part of the N electrons actually escape from the surface of the metal. A simple calculation, allowing for the absorption of both light and electrons, shows that the number N_1 which escape from unit area in unit time is

$$N_1 = \frac{\beta I}{a + \beta} N \qquad . \qquad . \qquad . \quad (38)$$

The definition of the absorption coefficient a as used in these calculations is not that usually given. The meaning of a may be deduced from the statement, that of all the electrons ejected from atoms in a plane slab of infinitesimal thickness perpendicular to the radiation, the proportion e^{-ax} reach a parallel plane distant x from the slab.

If

$$i_\nu d\nu = \frac{2}{c^2} \frac{h\nu^3 d\nu}{e^{h\nu/kT} - 1}, \qquad . \qquad . \qquad . \quad (39)$$

8

using Planck's notation, the corresponding number $N_2 d\nu$ emitted by the isotropic radiation inside the material, $i_\nu d\nu$ of frequency between ν and $\nu + d\nu$, is, by a similar calculation to that which leads to (38)

$$N_2 d\nu = \frac{4\pi N \beta i_\nu d\nu}{a} \qquad . \qquad . \qquad . \qquad (40)$$

whence we find for the total auto-photoelectric emission, after eliminating N by means of (38),

$$N_2 = \int_0^\infty N_2 d\nu = \frac{8\pi}{c^2} \int_0^\infty \frac{a + \beta}{a} \frac{N_1}{I} \frac{h\nu^3}{e^{\frac{h\nu}{kT}} - I} d\nu \quad . \quad (41).$$

For any particular temperature T, the integral on the right-hand side of (41) can be evaluated graphically if we know N_1/I and β/a for all frequencies. Values of N_1/I for all frequencies for which the factor

$$h\nu^3 / \left(e^{\frac{h\nu}{kT}} - I \right)$$

is appreciable have been given by Richardson and Rogers,[1] but much less is known about the values of a and β.[2] The only datum bearing on the value of these coefficients is an observation by Rubens and Ladenburg[3] who found that when ultra-violet light passed through a thin gold leaf the emission of electrons from the front side was 100 times as great as on the side of emergence, whereas the intensity of the incident light was 1000 times that of the emergent light. From these numbers Partzsch and Hallwachs[4] have calculated that for gold $a = 1\cdot03 \times 10^6$ cm.$^{-1}$ and $\beta = 0\cdot59 \times 10^6$ cm.$^{-1}$. Since gold and platinum do not differ much from one another in atomic weight and density, the value of the electron absorption coefficient a will probably be much the same for both metals. According to some data deduced by Drude[5] from katoptric measurements with sodium light β for platinum is greater than for gold in the ratio $4\cdot26/2\cdot82$. If this ratio holds also

[1] " Phil. Mag.," loc. cit.
[2] Cf., however, O. W. Richardson, *ibid.*, Vol. XXXI, p. 149 (1916).
[3] " Verh. der Deutsch. Physik. Ges.," Jahrg. 9, p. 749 (1907).
[4] " Ann. der Physik," Vol. XLI, p. 269 (1913).
[5] " Lehrbuch der Optik," 1st edition, p. 338.

for the rays which are photoelectrically active the value of β for platinum would become 0.89×10^6 cm.$^{-1}$ instead of 0.59×10^6 cm.$^{-1}$. Thus, with the data available the best estimate that we can make is that

$$(a + \beta)/a = 1.86 \qquad . \qquad . \qquad . \quad (42)$$

This is to be taken as a rough average value over the wave-lengths which are active photoelectrically. As the values for particular wave-lengths are not known the fraction (42) has been taken outside the integral in computing (41). From the known values of h and k and from the values of N_1/I given by Richardson and Rogers it appears, on evaluating (41), that the auto-photoelectric saturation current from platinum at $T = 2000°$ K. is

$$N_2 e = 5 \times 10^{-11} \text{ amp./cm.}^2 \qquad . \qquad . \quad (43)$$

In this calculation allowance has been made for the loss of light by reflexion from the platinum surface in the photoelectric measurements. On the other hand, the intensity of the electromagnetic radiation in the hot metal has been taken to be the same as that of the corresponding black body radiation in free space. This assumption is erroneous, but not likely to alter the order of magnitude of the final result.

The value (43) of the auto-photoelectric current is much smaller than the values of the thermionic currents ordinarily observed. Thus at $2000°$ K. Langmuir[1] gives the following values for the thermionic current densities for a number of elements which include platinum :—

Element $\rightarrow$	W.	Ta.	Mo.	Pt.	C.
Thermionic Current $\rightarrow$ (amps. per sq. cm.)	3×10^{-3}	7×10^{-3}	13×10^{-3}	6×10^{-4}	10^{-3}

The smallest thermionic currents ever recorded from platinum in the neighbourhood of $2000°$ K. are those observed by H. A. Wilson[2] with well-oxidized wires. He found 4×10^{-8} amp./cm.^{2} at $1686°$ C. which corresponds to nearly 10^{-7} amp./cm.^{2} at $2000°$ K.

We are thus led to the conclusion that the auto-photoelectric emission gives rise to an insignificant portion only of the

[1] "Phys. Rev.," Vol. II, p. 484 (1913).
[2] "Phil. Trans., A.," Vol. CCII, p. 243 (1903).

observed thermionic currents. This conclusion is only to be regarded as one to which the evidence at present available points with considerable probability. It cannot be held to be established beyond the possibility of a doubt, on account of the uncertainty involved in some of the assumptions underlying the calculations. In view of the importance of the subject it is very desirable that this question should be settled quite definitely ; but it is questionable whether any considerable advance on the present calculations can be effected without a material extension of our knowledge of the conditions underlying photoelectric action. Some further advance in this direction, which tends to strengthen the conclusion reached above, has recently been made by K. T. Compton and L. W. Ross.[1]

[1] " Phys. Rev.," Vol. XIII, p. 374 (1919).

CHAPTER IV.

THE EFFECT OF GASES ON THE EMISSION OF ELECTRONS.

1. THE EMISSION FROM PLATINUM IN AN ATMOSPHERE OF HYDROGEN.

WE have seen that the emission from a well-purified and oxidized platinum wire is affected very little by the presence of an atmosphere of a number of the commoner gases at various low pressures. Hydrogen, on the other hand, as H. A. Wilson found, has a very marked effect in stimulating the emission from this metal. The emission from platinum in an atmosphere of hydrogen has been the subject of a large number of experiments both by H. A. Wilson [1] and also by the writer.[2] The observations are in some minor respects not entirely harmonious, a result which probably arises from the complexity of the phenomena in detail and from the difficulty of distinguishing the relative importance of different influencing factors. The following discussion is confined to the more important points, about which the agreement is substantial.

In the first place, even in hydrogen, the emission at constant pressure is found to follow the formula $i = AT^{\frac{1}{2}}e^{-b/T}$. The constants A and b have, however, different values from those which characterize the emission in other gases and in a vacuum. As regards the constancy of A and b they are independent of the temperature, but may be functions of the pressure p of the hydrogen.

Working with wires which had not been heated in hydrogen for long periods Wilson arrived at the following conclusions:

---

[1] "Phil. Trans., A.," Vol. CCII, p. 263 (1903); *ibid.*, Vol. CCVIII, p. 247 (1908); "Roy. Soc. Proc., A.," Vol. LXXXII, p. 71 (1909); "Electrical Properties of Flames," etc., p. 16 (London, 1912).

[2] "Phil. Trans., A.," Vol. CCVII, p. 1 (1906).

When a wire, whose temperature was maintained constant, was allowed to remain for some time in hydrogen at different pressures the emission assumed steady values which were found to be governed by the formula

$$i = Bp^z, \qquad . \qquad . \qquad . \qquad . \qquad (1)$$

where B and z are independent of the pressure but depend on the temperature. Throughout the range of temperature used z was always between 0·5 and 1·0 and increased as the temperature diminished. When a change from one pressure to another was made, the emission did not immediately assume its final value but changed gradually from the value characteristic of the original pressure to that proper to the final pressure. A similar time lag in the value of the emission was observed when the temperature was changed at constant pressure. These effects are at once accounted for if it is admitted that the emission is determined not directly by the pressure of the external hydrogen but by the amount of hydrogen which is dissolved in the wire; since it is an established fact that hydrogen requires an appreciable time to diffuse through platinum. Since the emission at constant temperature diminishes with time, after passing from a lower temperature, and increases with time, after passing from a higher temperature, it is necessary to suppose that the equilibrium amount of dissolved hydrogen diminishes as the temperature rises. At least this must be the case under the particular conditions which Wilson[1] gives as an illustration, viz.: $p = 0·112$ mm. and T varying between 1284° C. and 1520°.C. The following numbers given by Wilson[2] indicate the way in which the current under 40 volts potential difference varies with the pressure at 1340° C. :—

Pressure (mms.).	Current (amps.).	Current ÷ Pressure.
760	3×10^{-3}	$3·9 \times 10^{-6}$
450	$1·5 \times 10^{-3}$	$3·3 \times 10^{-6}$
156	$2·2 \times 10^{-4}$	$1·41 \times 10^{-6}$
14	6×10^{-5}	$4·3 \times 10^{-6}$
0·11	4×10^{-6}	$36·0 \times 10^{-6}$
0·0013	5×10^{-8}	$39·0 \times 10^{-6}$

[1] "Phil. Trans., A.," Vol. CCII, p. 267 (1903).
[2] *Ibid.*, Vol. CCVIII, p. 255 (1908).

The effect of hydrogen on the constants A and b is shown in the following table taken from the same paper :—[1]

Gas.	Pressure (mms.).	A1.	b (Observed).	b (Calculated).
Air (1)	small	$7\cdot14 \times 10^{26}$	$7\cdot25 \times 10^4$	$7\cdot1 \times 10^4$
Air (2)	small	$4\cdot38 \times 10^{26}$	$6\cdot55 \times 10^4$	$6\cdot9 \times 10^4$
Hydrogen	$0\cdot0013$	$6\cdot25 \times 10^{24}$	$5\cdot5 \times 10^4$	$5\cdot5 \times 10^4$
Hydrogen	$0\cdot112$	$3\cdot13 \times 10^{23}$	$4\cdot5 \times 10^4$	$4\cdot52 \times 10^4$
Hydrogen	$133\cdot0$	$1\cdot25 \times 10^{21}$	$2\cdot8 \times 10^4$	$2\cdot7 \times 10^4$

The first value for air was given by a wire which had been subjected to the nitric-acid treatment previously described (p. 85) for twenty-four hours, the second only for one hour. The calculated values of b will be considered later.

The phenomena described above are quite different from those observed by the writer[2] under what appeared to be similar conditions. For example, the emission from a platinum wire in hydrogen was found to be practically constant whilst the pressure of the gas was reduced from 1 mm. to 0·001 mm. It was also found that a wire which had been saturated with hydrogen at a relatively high pressure and temperature could, after the pressure had been reduced, give off very considerable quantities of gas, presumably hydrogen, without the emission being much affected thereby. Similar results have since been observed by Wilson.[3] In one of his experiments there was very little variation in the emission when the pressure of the hydrogen was reduced from 200 mms. to 0·001 mm. He also confirmed the result that a wire which has been saturated with hydrogen may be heated in a good vacuum until it ceases to evolve gas, and retain the high power of emission previously conferred by the immersion in hydrogen, but he found that a wire under these circumstances still contains a large quantity of hydrogen which it is capable of retaining with great obstinacy. The presence of hydrogen in the wire under these conditions was established both by heating it in oxygen and measuring the diminution in pressure when the water formed was absorbed, and also by measuring the resistance of the platinum, which is affected by the presence of absorbed hydrogen.

It is clear from the facts which have been described that

[1] " Phil. Trans., A.," Vol. CCVIII, p. 251 (1908).
[2] *Ibid.*, Vol. CCVII, p. 51 (1906).
[3] *Ibid.*, Vol. CCVIII, p. 255 (1908).

the emission from platinum in an atmosphere of hydrogen shows two quite distinct types of behaviour, under conditions which at first sight appear to be identical. The cause of this difference was investigated by Wilson who found that the condition in which the emission was sensitive to changes in the pressure of the hydrogen occurred only with "fresh" wires, that is to say, with wires which had not been heated in hydrogen for any considerable length of time. The condition of insensitiveness to change of pressure, on the other hand, was found to be characteristic of wires which had been subjected to continued heating in hydrogen. Such wires may, for brevity, be described by the term "old". Wilson has pointed out that the observed facts are consistent with the view that in a fresh wire the hydrogen exists in a state of solution, whereas in an old wire most of it is present in the form of a compound which is formed with extreme slowness. The essential difference between solution and chemical combination lies in the fact that the amount of gas dissolved is a continuous function of the external pressure, whereas the amount chemically combined, once the reaction has been completed, is constant, if the external pressure exceeds the dissociation pressure, or zero if the pressure is below that value. Since the amount of hydrogen present in the wire is held to determine the value of the emission at a given temperature, on this view the emission will only be a continuous function of the pressure of the external hydrogen with fresh wires, in which the gas is present in the dissolved state. The dissociation pressure of the hydrogen compound must be very small (under 0·001 mm.) at the temperature of the experiments referred to (up to 1400° C.), since the large emission from an old wire could not be removed by pumping. However, dissociation pressures vary rapidly with temperature, and Wilson has found that the large emission from an old wire can be "pumped out" if the temperature is raised to about 1700° C. This view is also substantiated by the fact that the large emission from an old wire can be "burnt out" almost instantaneously in an atmosphere of oxygen at much lower temperatures; since the pressure of hydrogen in equilibrium with water vapour and its dissociation products under the conditions of such an experiment is very low.   It is, however, possible that the burning out process contaminates the platinum surface and reduces the emission to a value less than that which would characterize a clean platinum surface.

The writer [1] noticed a peculiar susceptibility to changes in the electric field in examining the emission from platinum wires in an atmosphere of hydrogen at pressures comparable with 1 mm.   The currents showed approximate saturation with potential differences of about 10 volts, and the increase due to impact ionization began to be observable definitely at about 20 volts.   Over a considerably higher range of potential than this the current was found to be a definite function of the applied voltage; but when the voltage exceeded 200, the current rapidly fell away from its initial value with lapse of time. Thus at 1·77 mm. pressure, and 1084° C., a steady current of 147×10^{-8} amp. was obtained with a potential difference of 19 volts ; on raising the voltage to 286 the current readings had the following values at the times stated :—

Time (minutes)	2	3	5	7	10	13
Current ($1 = 10^{-8}$ amp.)	62	44	37	33	28·5	26

If the high potential is maintained, the current goes on dropping, but at a diminishing rate, for several hours.   If the high potential difference is removed after most of the drop in the current has taken place and is replaced by a low voltage, the observed phenomena are quite different and rather surprising. The current with the low voltage is small at first and remains practically constant and very steady for a considerable interval. It then begins to increase, slowly at first, then more rapidly, then slowly again and finally attains a constant value which is much higher than the steady value finally attained under the high voltage applied previously.   Thus, after the numbers in the table above had been obtained, when the wire was giving a current of 26×10^{-8} amp. after exposure to 286 volts for thirteen minutes, the potential was suddenly changed to 80 volts.   The current at once dropped to 7×10^{-8} amp. and remained at this value for several minutes.   It then began to rise, the rate of increase being most rapid after the lapse of

[1] " Phil. Trans., A.," Vol. CCVII, p. 46 (1906).

100 minutes. Fifty minutes later the current had attained the practically constant value of 220 × 10^{-8} amp. This is more than ten times the value of the steady current which would have been attained if the potential difference of 286 volts had been left on indefinitely.

These effects did not occur either at very low pressures (< 0·1 mm.) or at high pressures (200 mm.) or when the wire was charged positively. The conditions under which they were found to occur are those under which the platinum would be subjected to an energetic bombardment by positive ions arising from impact ionization. It seems natural to interpret the falling off of the current under the application of high voltages to the destruction [1] by this bombardment of a structure at the surface of the platinum which facilitates the escape of the electrons. This structure will begin to form again as soon as low voltage conditions are restored, and the bombardment becomes ineffective. It is doubtless responsible, at least in part, for the increased emission in a hydrogen atmosphere.

An effect which is probably related to this has been observed by Wilson.[2] After the large emission from a wire saturated with hydrogen has been removed by heating in oxygen the wire appears to recover its lost power very slowly when subsequently heated in a hydrogen atmosphere; but the rate of recovery is very greatly increased, if the wire forms one of the electrodes in the passage of a luminous discharge in hydrogen at a low pressure.

Platinum is not the only metal from which the emission of electrons is increased by immersion in an atmosphere of hydrogen. Similar effects have been noted by Wilson[3] with palladium and by J. J. Thomson[4] with sodium. All three elements are notable for their power of dissolving and combining with hydrogen. On the other hand, Langmuir[5] found that hydrogen caused an enormous reduction in the emission

[1] Or conversely to the formation of a structure which retards their escape, but this hypothesis seems less probable.

[2] "Phil. Trans., A.," Vol. CCVIII, p. 265 (1908).

[3] *Ibid.*, Vol. CCII, p. 271 (1903).

[4] "Conduction of Electricity through Gases," 2nd edition, p. 203.

[5] "Phys. Rev.," Vol. II, p. 463 (1913)

from tungsten, an effect which he attributes, however, to the action of water vapour formed by secondary actions, rather than to the direct action of the hydrogen itself.

Theory of the Emission from Fresh Platinum Wires in Hydrogen.

The essential features of the emission from fresh platinum wires in an atmosphere of hydrogen may be summarized in the two following statements :—[1]

(1) The emission at any constant temperature is very closely proportional to p^z, where the index z is a proper fraction whose value diminishes with rising temperature; and (2) under constant pressure the emission at different temperatures is governed by a formula of the type $i = AT^{\frac{1}{2}} e^{-b/T}$, where, however, the constants A and b are functions of the pressure of the hydrogen, and their values are, in general, quite different from those of the corresponding quantities appropriate to the emission from platinum wires in a vacuum.

These laws, including the actual changes in the values of the constants A and b, must be intimately connected with the contact potential differences between pure platinum and platinum immersed in an atmosphere of hydrogen. This follows from considerations similar to those pointed out already in Chapter III, page 91. The presence of hydrogen will alter the change of energy which accompanies the escape of an electron, as well as change the contact potential. These two effects are closely connected together but they are not identical, and the change in the contact potential is more directly related to the change in the emission constants than is the interfacial change in energy. The distinction is an important one and one which is particularly liable to confusion ; so that it is worth while to devote a little consideration to the purpose of making it clearer.[2] To do so it is necessary to refer to certain matters of theory which were touched on at the beginning of Chapter II, where, by making use of the laws of thermodynamics, certain relations were deduced between

[1] Cf. H. A. Wilson, " Phil. Trans., A.," Vol. CCVIII, p. 248 (1908).

[2] For a fuller discussion cf. O. W. Richardson, " Roy. Soc. Proc., A.," Vol. XCI, p. 524 (1915).

the equilibrium concentrations of the external electrons, on the one hand, and the energy change ϕ and the specific heat of electricity in the hot body, on the other. When gases are present the proofs of these relations which have been given require modification. A metal in the presence of a gas is a much more complex affair than a pure metal; but, to a first approximation, it can be represented as a piece of pure metal covered with a layer of contamination of definite composition and of finite thickness. If the proofs are modified, so as to suit such a structure and to allow for the presence of the gas, it still appears that

$$n = A e^{\int^T \frac{\phi}{kT^2} dT} \quad . \qquad . \qquad . \qquad . \qquad (7')$$

where A is independent of T and ϕ is now the energy change when an electron passes through the outer surface of the contaminated layer. Obviously ϕ is not a quantity which is directly accessible to measurement. In carrying out the cycle on page 32 involving the transference of electrons between two metals at different temperatures we have now to include the work η which corresponds to the Peltier effect at the interface between the pure and uncontaminated surfaces and by so doing we obtain, instead of equation (13) (Chapter II) and after taking out the logarithms and replacing p by nkT,

$$n = C T^{\frac{1}{\gamma-1}} e^{-\left\{ \frac{\phi+\eta}{kT} + \frac{\epsilon}{k} \int^T \frac{\sigma}{T} dT \right\}} \quad . \qquad . \qquad (13')$$

where C is a universal constant. By making use of a known result in thermoelectricity this may be replaced by

$$n = C T^{\frac{1}{\gamma-1}} e^{-\left\{ \frac{\phi}{kT} + \frac{\epsilon}{k} \int^T \frac{\sigma_1}{T} dT \right\}} \quad . \qquad . \qquad (13'')$$

if σ_1 is the specific heat of electricity in the contaminated material. Thus (13) of Chapter II is still valid, as it should be, if we keep to a single definite substance. If ϕ', η' ($= 0$), and n' are the values for the pure metal,

$$n' = C T^{\frac{1}{\gamma-1}} e^{-\left\{ \frac{\phi'+\eta'}{kT} + \frac{\epsilon}{k} \int^T \frac{\sigma}{T} dT \right\}}$$

and

$$n' = n e^{-\left\{ \frac{\phi'+\eta'-(\phi+\eta)}{kT} \right\}}$$

$$= n e^{\frac{-\epsilon V_1}{kT}} \quad : \qquad . \qquad . \qquad . \qquad (34')$$

where[1] V_1 is the contact difference of potential between the pure and the contaminated materials. This equation is the same as equation (34) of Chapter II, and may be obtained in a similar way to that followed on page 42. Since, if we neglect electron reflexion, the saturation currents i, i' at a given temperature are in the same ratio as the corresponding equilibrium electron concentrations n, n', we may replace (34') by

$$i' = ie^{-\epsilon V_1/kT} . \qquad . \qquad . \qquad (34'')$$

We know from experiment that for the pure metals

$$i' = A'T^\lambda e^{-\omega'/kT} \qquad . \qquad . \qquad . \qquad (2)$$

where A', ω' ($= kb'$) and λ are independent of T, λ being comparable with unity. Now ϵV_1 will depend on the temperature and on the pressure p of the hydrogen. To indicate this, let us denote its value by kTf (p, T) where f is a function whose nature we shall seek to determine. Then the expression for the current when hydrogen is present will become, from (2) and (34''),

$$i = A'T^\lambda e^{-\frac{\omega}{kT}+f(p,\,T)} \qquad . \qquad . \qquad (3)$$

Since (3) has to become identical with (2) when $p = 0$, $f(p, T)$ must approach zero as p approaches zero. Again, the experimental results show that (3) as a function of T has the same form as (2), but that the new constants corresponding to A' and ω' are now functions of p but not of T. This requirement is satisfied if f (p, T) is of the form $F_1(p) + \dfrac{1}{T} F_2(p)$, where F_1 and F_2 are functions of p only; for we can then write (3) in the form

$$i = A'e^{F_1(p)}T^\lambda e^{-[\omega - kF_2(p)]/kT}, \qquad . \qquad . \qquad (4)$$

the new values of the constants at pressure p being

$$A'' = A'e^{F_1(p)} \text{ and } \omega'' = \omega' - kF_2(p) . \qquad . \qquad (5)$$

Again the law of variation of emission with pressure at constant temperature requires that (3) should approach very closely to

$$i = Bp^x \qquad . \qquad . \qquad . \qquad (1)$$

---

[1] " Electron Theory of Matter," p. 456 (Cambridge, 1914).

where B and z are functions of T only, z diminishing as T increases and lying between $\frac{1}{2}$ and 1 over the range of temperatures covered in the experiments. It is clear, however, that (1) cannot be the complete function, otherwise i would vanish when $p = o$, whereas we know that i then reduces to the value characteristic of a vacuum. But if we replace (1) by

$$i = B(1 + ap^c)^z \qquad . \qquad . \qquad . \qquad (6)$$

where a and c are constants and ap^c is large compared with unity at all pressures which are measurable, both requirements are satisfied. By comparing with equation (4) we then get, omitting all the terms which do not involve p,

$$(1 + ap^c)^{\frac{z}{c}} = e^{F_1(p) + \frac{1}{T}F_2(p)}$$

or

$$\frac{z}{c} \log (1 + ap^c) \equiv F_1(p) + \frac{1}{T}F_2(p) \qquad . \qquad . \qquad (7)$$

Since z is independent of p, and since F_1 and F_2 are independent of T and are not constant, the only values of these functions which are compatible with (7) are

$$F_1(p) = \frac{a'}{c} \log (1 + ap^c) . \qquad . \qquad . \qquad (8)$$

$$F_2(p) = \frac{a}{c} \log (1 + ap^c) . \qquad . \qquad . \qquad (9)$$

$$z = a/T + a' \qquad . \qquad . \qquad . \qquad (10)$$

Wilson[1] has shown that, if $a = 1\cdot27 \times 10^4$, $c = 0\cdot73$, $a = 2\cdot4 \times 10^8$, and $a' = -c$, the experimental results are accounted for with very considerable approximation over the whole of the wide ranges of temperature and pressure which have been covered in the quite considerable number of investigations available. Substituting the new expression for the contact potential in equation (2) we have for the current at pressure p and temperature T°K

$$i = A'T^\lambda e^{-\left[-kT \log (1 + apc)\left(\frac{a}{cT} - 1\right)\right]/kT}$$

$$= A'(1 + ap^c)^{\left(\frac{a}{cT} - 1\right)}T^\lambda e^{-\omega' k/T} \qquad . \qquad . \qquad (11)$$

[1] "Electrical Properties of Flames," etc., p. 20.

$$= A'(1 + ap^c)^{-1} T^\lambda e^{-\dfrac{\omega' - k\frac{a}{c}\log(1+ap^c)}{kT}} \qquad . \quad (12)$$

From (11) we see that when the temperature is constant and the pressure varies

$$i \propto (1 + ap^c)^{\left(\frac{a}{cT}-1\right)} \qquad . \qquad . \qquad . \quad (13)$$

$$\propto p^z, \text{ very nearly, if } z = \frac{a}{T} - c,$$

since unity is very small compared with ap^c for the pressures in question. The extent to which this result is satisfied may be illustrated by means of the following numbers given by Wilson :—[1]

Pressure (mms. of mercury).	i (arbitrary units).	$i \div$ right-hand side of (13).
0·0006	10	2·65
0·0015	20	2·70
0·0033	40	3·01
0·0053	50	2·65
0·0080	75	2·92
0·0140	110	2·85

From (12) we see that i still satisfies an equation of the type

$$i = A'' T^\lambda e^{-\omega''/kT}, \qquad . \qquad . \qquad . \quad (14)$$

but that the constants A'' and ω'' have new values, being given by

$$A'' = A'(1 + ap^c)^{-1} \qquad . \qquad . \qquad . \quad (15)$$

$$\text{and } \omega'' = \omega' - k\frac{a}{c}\log(1 + ap^c) \qquad . \qquad . \quad (16)$$

$$= \omega' - k\frac{a}{c}\log \frac{A'}{A''} \cdot \qquad . \qquad . \quad (17)$$

We have seen that it is very difficult to determine the correct values of A' or A'', but that (15) is very near to being satisfied is shown by the fact that when p is changed from 0·0013 mm. to 133·0 mm., A'' diminishes regularly till it becomes only about $1/5000$ of its initial value, whereas $A' = A'' \times (1 + ap^c)$ only changes in an irregular manner and shows an extreme variation of a factor of 3. Again from (17) we see that the values of ω' can be calculated from the value ω for a

---

[1] " Electrical Properties of Flames," etc., p. 16.

vacuum if we know the values of A′ for a vacuum and of A″ at different pressures. The values of $b = \omega''/k$ calculated in this way are given in the table on p. 119 and show an excellent agreement with the observed values.

In the last chapter, p. 102, in discussing Martyn's experiments on the emission from platinum and lime in air and in hydrogen, it was pointed out that the effect of the gas on the platinum could be accounted for by supposing that the contact difference of potential between pure platinum and platinum in an atmosphere of hydrogen was equal to $kT \log p_1/p_3$, where p_1 and p_3 are the pressures of the *positive* hydrogen ions inside and outside the metal respectively. This view becomes identical with the one now under consideration if we suppose that

$$p_1/p_3 = (1 + ap^c)^{\left(\frac{a}{cT}-1\right)}. \qquad . \qquad . \quad (18)$$

where p is the pressure of the external hydrogen gas. So far as the writer is able to judge there is not enough known about the phenomena attending the solution of hydrogen in platinum to enable one to determine whether a formula such as (18) is likely to approximate to the truth or not. It is clear, however, that the right-hand side of (18) varies with temperature qualitatively in the way to be expected. At low temperatures the dissociation (and ionization) of the external hydrogen is small, both absolutely and also, most probably, in comparison with that of the internal hydrogen. Since there is evidence that positive hydrogen ions are formed with less expenditure of energy inside the metal than outside, it follows that the external ionization will increase more rapidly with rising temperature than the internal ionization. Since p_1 and p_3 are proportional to the respective numbers of positive hydrogen ions per cubic centimetre, it follows that the right-hand side of (18) should have a large value at low temperatures which should diminish very much as the temperature rises. This is found to be the case. Thus when $p = 1$ mm. the right-hand side of (18) is equal to 3×10^8 at 800° C. and falls to $2\cdot4 \times 10^2$ at 1800° C.

Another point to be borne in mind is that the right-hand side of (18) has been obtained in a manner which is largely empirical and, like all functions which are derived in this way, is subject to the liability that it is not the true function but merely one which simulates the mathematical behaviour of the true function. If this is the case, however, it is necessary that the correspondence should be very close indeed. For the formulæ give results very near to the truth, not only over such limited ranges of pressure as those illustrated in the table on p. 127, but over all the very extended range of temperature and pressure which has been examined by various experimenters. Thus if $i = i_0$ when $p = 0$, (13) may be written

$$i/i_0 = a(1 + ap^c)^{\left(\frac{a}{c\mathrm{T}} - \right)}. \qquad . \qquad . \qquad (19)$$

At 900° C. at 26 mms. pressure the writer[1] found $i/i_0 = 4 \times 10^8$. The calculated value given by the right-hand side of (19) is 2×10^9. If c is put equal to 0·78 instead of 0·73 the right-hand side of (19) becomes 3×10^8 instead of 2×10^9. Thus the apparent disagreement may be attributed to the uncertainty in the precise value of the constants. Again at 1570° C. and 760 mms. Martyn[2] found $i/i_0 = 4\cdot4 \times 10^4$, the calculated value from (19) being $6\cdot5 \times 10^4$. At 1340° C. and 0·0013 mm. Wilson[3] found $i/i_0 = 170$, the calculated value being 122. Thus the values given by (19) are very close to the actual values over the range of pressure from 0 to 760 mms. and of temperature from 900° to 1575° C.

By comparing (18) and (19) we see that

$$\xi = i/i_0 = (1 + ap^c)^{\left(\frac{a}{c\mathrm{T}} - 1\right)} = p_1/p_3 . \qquad . \qquad (20)$$

so that on the hypothesis under immediate consideration the ratio ξ, of the saturation current at pressure p to that at pressure zero, is equal to the ratio of the pressure (or concentration) of the internal to the pressure (or concentration) of the external positive hydrogen ions, in the presence of platinum subject to a total external hydrogen pressure p. The whole subject is

[1] " Phil. Trans., A.," Vol. CCVII, p. 45 (1906).
[2] " Phil. Mag.," Vol. XIV, p. 306 (1907).
[3] " Electrical Properties of Flames," etc., p. 21.

well worth further investigation to see if this result is substantiated by more complete experimental knowledge.

It is evident from the argument on pages 42, 87, and 125 that the change in A′ caused by hydrogen may be attributed to the occurrence in the expression for the contact potential difference, between the metal in an atmosphere of hydrogen and the pure metal, of a term which is proportional to the temperature. In other words, the change in A′ may be considered to arise from the temperature variation of the changed value of ω′. We also notice that any change in the observed value of ω′ will arise only from that part of the expression for the contact potential which is independent of T. Thus if for any given value ω_1 of the work which corresponds to the contact potential difference we are able to calculate its variation with temperature we ought to be in a position to calculate the changed value of A′ corresponding to ω_1.

The problem has been attacked from this point of view by H. A. Wilson,[1] who calculates the whole work done in passing from a point in the interior of the pure metal to the outside. This amount of work includes both that which corresponds to the Peltier effect at the interface and the change of energy at the outer surface in the treatment given above, and is therefore equivalent to the contact potential difference. Wilson supposes that the work necessary for the escape of an electron arises from the presence of an electrical double layer of thickness t, and negatively charged on the outside, at the surface of the metal. Whether such a layer really exists or not, its imaginary presence will give rise to effects in many respects identical with, and in others similar to, those arising from the actual mechanism which causes the origin of the contact potential difference. It is to be remembered, however, that the conception is an artificial one whose chief attraction lies in its amenability to calculation. If the double layer consists simply of charges, of surface density $\pm \sigma$, at distance t apart, the work done in taking a change ϵ across it is $4\pi\sigma t\epsilon$. Wilson supposes that σ and t are independent of temperature, so that

---

[1] "Phil. Trans., A.," Vol. CCVIII, p. 268 (1908); "Roy. Soc. Proc., A.," Vol. LXXXII, p. 71 (1909); "Electrical Properties of Flames," etc., p. 22.

their values in any particular case determine the zero-temperature value of the work in question. The results indicate that *t* is also unchanged when hydrogen is present and that the effect of hydrogen may be interpreted as causing a reduction in the value of σ. The variation of the work with temperature is supposed to arise from the diffusion of the electrons into the surface layer owing to their heat motion. This diffusion will increase with rising temperature and will increase the effective strength of the double layer; so that the work necessary for an electron to escape will increase as the temperature increases. The diffusion will also be greater at a given temperature the smaller the value of σ; so that the temperature coefficient when hydrogen is present will be greater than the temperature coefficient in its absence. The actual calculations involve a consideration of the equilibrium of the electrons in the double layer and are somewhat complicated. For them the reader may be referred to Professor Wilson's book, "The Electrical Properties of Flames and of Incandescent Solids," p. 23. As a result, if ω_1 is the value of the work under consideration, the calculations give the following values of $2\omega_1/k$ at temperature T°K.

Gas.	Pressure.	$2\omega_1/k$.
Air	—	145,000 + 2·35 T
H_2	0·0013	110,000 + 11·83 T
H_2	0·112	90,000 + 17·82 T
H_2	133·0	56,000 + 28·86 T

These numbers lead to values of A' and ω' in hydrogen at different pressures in close agreement with the experimental values already considered. It is also found that if the equations are solved for *t* the different values of A' and ω' always lead to values between $t = 5\cdot6 \times 10^{-8}$ cm. and $10\cdot7 \times 10^{-8}$ cm. These can be considered to be constant within the limits of error. The values are also in agreement with the value of the thickness of the double layer calculated from the polarization capacity of platinum polarized with hydrogen by the electrolysis of dilute sulphuric acid.

It is interesting to observe that the value of the temperature coefficient of ω_1/k in air or a vacuum given by these calculations is very close to that of ϕ given by the theory in

9 *

Chapter II. If the temperature variation of ω_1/k were made equal to that of ϕ we see, by comparison with equation (16), Chapter II, page 35, that the value of $2\omega_1/k$ would be changed only from $145,000 + 2\cdot35$ T to $145,000 + 3$T.

THE EFFECT OF GASES ON THE EMISSION FROM TUNGSTEN.

Tungsten possesses a number of notable advantages for the purpose of experiments on the emission of electrons from hot bodies. It is the most refractory material known, melting at 3270° C., and its volatility is low even at the highest temperatures. Thus it can be heated, without any considerable loss by evaporation, for comparatively long periods at temperatures so high that all known impurities are driven out of it. At these high temperatures enormous electron currents may be obtained, the only limits being set by the heating current required to fuse the wire which is the source of supply, and by the potential difference required to overcome the mutual repulsion of the emitted electrons. For example, the writer has observed a thermionic leakage of 0·4 amp. from a fine filament which required 0·8 amp. heating current. In this case the thermionic current density amounted to 4 amp. per sq. cm. The large currents from tungsten are absolutely steady when attention is paid to the proper preparation of the tubes (see p. 13) and are very suitable for exact work. Moreover, owing to the importance of tungsten as a material for lamp filaments, its electrical and radiating properties at high temperatures have been very thoroughly studied. Finally it acts as a self-purifying agent by attacking all except the inert gases, forming compounds which are then volatilized on to the walls of the tube.

The effect of different gases on the emission from tungsten at about 2000° C. has been investigated by Langmuir.[1] In these experiments the thermionic currents were larger than those usually dealt with, being for the most part in the

[1] "Phys. Rev.," Vol. II, p. 450 (1913); "Phys. Zeits.," Jahrg. 15, p. 516 (1914).

neighbourhood of 1-20 milliamperes. The difficulty in attaining saturation owing to the mutual repulsion of the electrons was therefore generally an important factor (see p. 68). The gases experimented with include hydrogen, water vapour, oxygen, nitrogen, and argon. When saturation was attained the currents were always found to be capable of representation by the formula $i = AT^{\frac{1}{2}}e^{-b/T}$. The constants A and b have, however, in general different values in different gases. The values also are probably different for the same gas at different pressures. Very small amounts of gas were found to cause very large changes in the values of the constants. The values of the constants under good vacuum conditions varied very little in different experiments either with the same or with different tubes. The following numbers are cited as the results of separate determinations for this case :—

$$A = 6 \cdot 6 \times 10^{16} \text{ e.s.u. per cm.}^2 \qquad b = 5 \cdot 58 \times 10^4 \text{ degrees C.}$$
$$A = 10 \cdot 2 \times 10^{16} \qquad\qquad b = 5 \cdot 55 \times 10^4$$
$$A = 7 \cdot 08 \times 10^{16} \qquad\qquad b = 5 \cdot 25 \times 10^4$$

It was found that the values of the constants were not appreciably altered by the presence of argon. The saturation currents in this gas have, so far as can be ascertained, the same value as in a vacuum. The only effect of the argon when present in small quantity is to facilitate the attainment of saturation through the action of the positive ions, formed by impact ionization, in reducing the effect of the mutual repulsion of the electrons. This result is of great importance. No doubt when the pressure of the argon is appreciable the current will be magnified owing to ionization by collisions, but the effect would not be of importance in the experiments now under consideration owing to the low pressures (up to 0·002 mm.) of the argon used. Mercury vapour,[1] the other inert gases of the argon group,[2] and hydrogen,[2] when pure, also behave like argon in these respects.

All the other gases tested were found to increase the values of *both* the constants, as is shown by the following numbers :—

---

[1] Richardson, " Phil. Mag.," Vol. XXVI, p. 347 (1913).

[2] Langmuir, " Trans. Amer. Electrochem. Soc.," p. 352 (1916).

Gas.	Pressure (mm.).	A.	b.
Vacuum	0·00007	$10·2 \times 10^{16}$	$5·55 \times 10^4$
H_2	0·012	$1·62 \times 10^{21}$	$8·25 \times 10^4$
H_2	0·0005	$1·29 \times 10^{22}$	$8·5 \times 10^4$
H_2	0·007	$2·28 \times 10^{28}$	$11·5 \times 10^4$
H_2	0·0017	$2·31 \times 10^{26}$	$10·5 \times 10^4$
O_2	—	$2·04 \times 10^{23}$	$9·43 \times 10^4$
N_2	0·002	$6·6 \times 10^{19}$	$7·32 \times 10^4$
N_2	—	5×10^{18}	$6·82 \times 10^4$

It is clear from the values for hydrogen that there is no relation between the magnitudes of A and b for this gas and the corresponding pressure. For this and other reasons Langmuir is inclined to attribute most of the change in the constants apparently caused by hydrogen to the effect of traces of water vapour either formed by it or introduced with it. In all cases the changes in the emission caused by the gases persisted for some time after the gas had been removed, showing that the effect was not directly due to an action between the filament and the external gas but to a semi-permanent change produced by the gas in the character of the tungsten surface. No doubt the precise determination of the constants in a given gas at a definite pressure is a difficult matter as all the gases except argon are " cleaned up " during the course of the experiment ; so that the pressure is continually diminishing in any given case. Of the three gases oxygen, nitrogen, and hydrogen, oxygen is absorbed most rapidly at about 2000°. The absorption of nitrogen appears to be an electro-chemical phenomenon which exhibits interesting effects. These will be considered later.

The magnitudes of the changes in A and b caused by different gases suggest that all these changes are due to a common cause, or, at any rate, that the mechanism of the action of the different gases is of such a nature as to possess important common features in the different cases. Thus in the table above the values of A and b always increase and diminish together. This is seen more clearly from Fig. 13 in which the values of $\log_{10}$ A are plotted against those of $b \times 10^4$. Fig. 13 contains all the data of the table and some others in addition. It will be observed that the values of $\log_{10}$ A, no matter what gas has given rise to them, are very near to satisfying a linear relation with the corresponding

values of b. It is very doubtful, owing to the time variations which must be occurring in many of these experiments,
whether the linear relation is not satisfied within the limits of
experimental uncertainty.

As with the similar effects observed with hydrogen and
platinum these effects may be considered in relation to the
contact potential difference between pure tungsten and tungsten which is contaminated with the gases under considera-

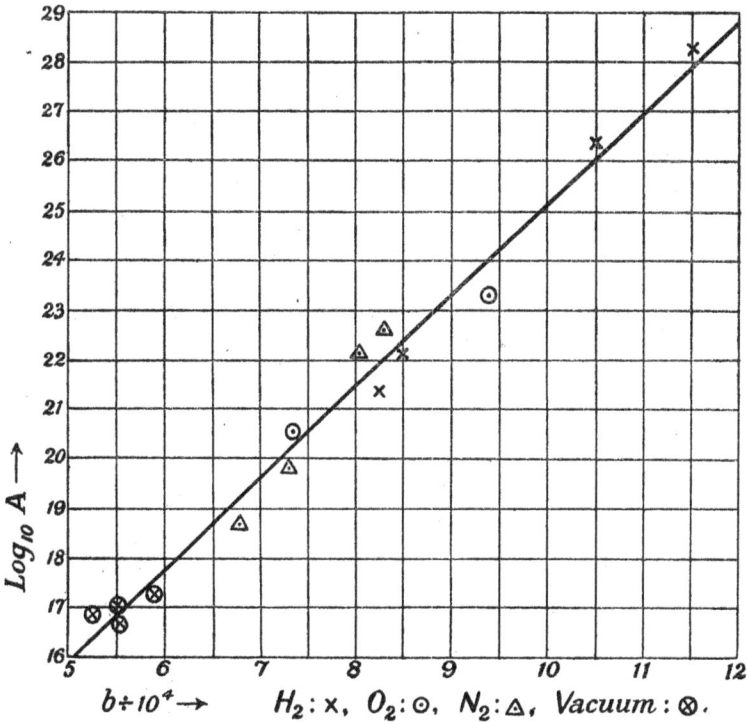

Fig. 13.

tion. In the absence of gas, i.e. in a thoroughly "cleaned
up" vacuum, let the saturation current be

$$i = A_0 T^{\frac{1}{2}} e^{-b_0/T}.$$

Suppose that the contact potential difference due to the
gas is $\dfrac{k}{\epsilon} (b_0' + \beta T)$ where k is Boltzmann's constant, ϵ is the
ionic charge, β is a constant, and b_0' may, in general, be a
function of T which contains no linear term in T, this term

being represented by βT.   The current when the gas has produced its effect will be represented by

$$i'' = A_0 T^{\frac{1}{2}} e^{-(b_0 + b_{0'} + \beta T)/T}$$
$$= A_0 e^{-\beta} T^{\frac{1}{2}} e^{-(b_0 + b_{0'})/T} . \qquad . \qquad . \qquad (21)$$

If the range of temperature is not too large this may still be represented by

$$i' = A T^{\frac{1}{2}} e^{-b/T} . \qquad . \qquad . \qquad (22)$$

with A and b constants, if b_0' does not vary too rapidly with T.   The values of A and b are

$$A = A_0 e^{-\beta} \qquad . \qquad . \qquad . \qquad (23)$$
$$b = b_0 + b_0' . \qquad . \qquad . \qquad (24)$$

Since A is always greater than A_0, β is negative, and from (23)

$$\gamma = -\beta = \log \frac{A}{A_0} \qquad . \qquad . \qquad (25)$$

If the linear relation indicated by Fig. 13 is really fulfilled, we have

$$b = b_0 + b_0' = c' \log \frac{A}{A_0} + c, \qquad . \qquad . \qquad (26)$$

where c and c' are constants.   Since $b_0' = 0$ when $A = A_0$, $c = b_0$.   Thus

$$b_0' = c' \log \frac{A}{A_0} = \gamma c' \qquad . \qquad . \qquad (27)$$

The contact potential difference due to the gas is thus equal to

$$kT \left(\frac{b_0'}{T} - \gamma \right) = \gamma k T \left(\frac{c'}{T} - 1 \right) \qquad . \qquad . \qquad (28)$$

In this equation k is a universal constant, c' is independent of T and has the same value for all the gases tested, γ is independent of T but is determined by the modification in the state of the surface caused by the gas.   Since for a given gas the only factor except T capable of controlling the state of the surface would appear to be the pressure of the gas, it would seem that ultimately γ must be a function only of the pressure of that gas which causes the change in question—γ may, however, be a different function of the pressure for each gas which gives rise to these effects.

The effect of these gases on the emission from tungsten shows a very close correspondence with the effect of hydrogen on the emission from platinum. Turning to the table on p. 119, remembering that A is proportional to A_1, we see that the effect on platinum of gradually increasing the pressure of the surrounding hydrogen, is to cause a corresponding series of changes in A and *b*. In these changes A and *b* always increase or diminish together. The chief difference between the effect of hydrogen on platinum and the effect of the various gases under consideration on tungsten is that, in the case of platinum the change from the normal is towards lower values of A and *b*, whereas, in the case of tungsten, the change is towards higher values. Moreover, it follows from the considerations brought forward on p. 125 *et seq.*, that the contact potential difference caused by the gases, considered as a function of pressure and temperature, is of the same form in both cases, the main difference being that the equivalent work is negative, corresponding to a more electropositive condition, in the case of hydrogen and platinum, whereas it is positive, corresponding to a more electronegative condition, when tungsten is contaminated by various gases. Thus from equation (11) p. 126, we see that the equivalent work in the case of hydrogen and platinum may be written

$$\delta w = -\left(\frac{a}{cT} - 1\right) kT \log\left(1 + ap^c\right) . \quad . \quad (29)$$

whereas from (28) the corresponding quantity for tungsten is

$$\delta w' = +\left(\frac{c'}{T} - 1\right) kT\gamma . \quad . \quad . \quad (30)$$

We have seen already that the quantity γ in (30) plays a similar part to the function $\log\left(1 + ap^c\right)$ in (29); so that the terms $\left(\frac{c'}{T} - 1\right)$ in (30) and $\left(\frac{a}{cT} - 1\right)$ in (29) are precisely comparable with one another. Moreover, the constants c' and a/c in these expressions have almost equal values. The data on p. 127 give

$$a/c = 3\cdot29 \times 10^3$$

whereas from Fig. 13

$$c' = 2\cdot56 \times 10^3.$$

A still closer agreement is obtained if all the known pairs of values of A and b for platinum are considered.[1]

It appears to be a legitimate inference from these results that the two effects under consideration are due to similar causes acting in opposite senses in the two cases. If in the case of platinum the cause lies in the difference of concentration of positive hydrogen ions inside and outside the metal, it would be natural to attribute the effects with tungsten to a difference in the concentration of negative ions, probably ions of the electronegative elements oxygen and nitrogen, inside and outside the surface layer. On the other hand, if the platinum effects arise from the action of positive hydrogen ions on a double layer at the surface, it would seem reasonable to ascribe the tungsten effects to the similar action of negative ions furnished by oxygen or nitrogen. It is to be remembered that the two hypotheses contrasted are not necessarily contradictory.

In an atmosphere of nitrogen at low pressures, Langmuir observed peculiar effects which were not exhibited by any of the other gases. At the lower temperatures tested, it was found that the electron currents in this gas were *larger* with small than with large potential differences. For example, the following currents in milliamperes per square centimetre were obtained under the conditions indicated in the table :—

Temperature °K.	220 Volts.	100 Volts.	220 Volts.
2045		0·34	0·29
2090		0·70	0·63
2140		1·50	1·29
2190	2·7	4·00	2·9
2250	6·3	4·9	7·0
2325	16·2	5·0	19·3
2390	21·0	5·0	20·0
Pressure of N_2 →	0·0015 mm.	0·0012 mm.	0·0012 mm.

The reason for this peculiar behaviour becomes clearer when the variation of current with applied potential difference at constant temperature is studied in nitrogen under different small pressures. The results of experiments of this character at 2100° K. in nitrogen at 0·00016 mm., 0·0010 mm.

[1] Cf. O. W. Richardson, " Roy. Soc. Proc., A.," Vol. LXXXIX, p. 524 (1915). In this article it is also shown that there are indications of a similar relation affecting the emission of positive ions from hot platinum. Cf. also p. 226.

and 0·0025 mm. pressure are shown in Fig. 14. The readings for a pressure of 0·00016 mm. approximate to those for good vacuum conditions. At low potentials they follow the full curve IV which represents the variation, with applied potential, of the current when limited by the mutual repulsion of the electrons (space-charge effect) as considered in Chapter III, p. 68. At higher potentials they leave this curve

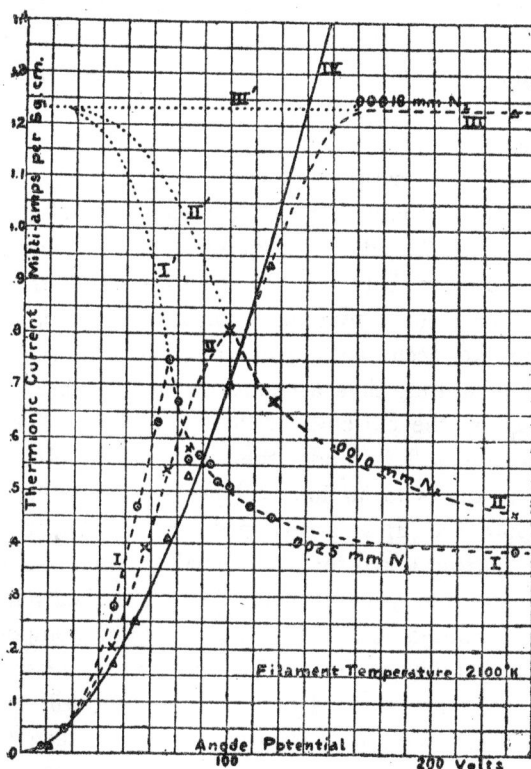

Fig. 14.

and ultimately fall on the horizontal dotted line III', which represents the saturation current under vacuum conditions at the temperature of the experiment. The dotted curve III, in fact, which passes through all the values at 0·00016 mm. is similar in a general way to the curves obtained in all the gases under consideration other than nitrogen. In nitrogen at higher pressures, however, the curves are quite different, as

is shown by I and II. Considering curve I, for example, we
see that it is coalescent with III and IV at potentials below
20 volts. At potentials between 20 and 75 volts the currents
given by I are larger than those given by III and IV. This
effect, which is similar to effects given by other gases, is attributed to the action of the positive ions produced by impact
in reducing the mutual repulsion of the electrons, and so permitting a nearer approach towards saturation. At 75 volts
the rise in the current with increasing potential suddenly
ceases, and is replaced by a fall which is most rapid at first
and then diminishes until a steady value of the current is
finally reached. This ultimate saturation current is much
smaller than the saturation value in a vacuum. The behaviour above 75 volts has so far been observed in nitrogen
only. It is attributed by Langmuir to the occurrence of a
chemical reaction between the tungsten and the positive
nitrogen ions formed by impact ionization. This reaction is
known not to occur with uncharged nitrogen molecules at
these temperatures, thus accounting for the absence of any
diminution of the currents at low potentials. The nitride
formed is supposed to hinder the escape of the electrons, and
as the rate of its formation will go on increasing with the
applied potential (up to a certain limit), the general course of
the factor cutting down the current will resemble a curve such
as I'. Thus the general character of the current-voltage
curves is accounted for. The amount of the compound ultimately formed will be greater the greater the pressure of the
nitrogen; so that the final saturation current will be reduced
as the nitrogen pressure is increased. This is seen to be the
case with the data given. Finally, as the temperature is
raised, the compound formed will evaporate more quickly
and so less of it will be retained on the filament. Thus this
effect should diminish at higher temperatures, as in fact is
found to be the case.

Returning to the general case of the effect of gases on
tungsten we have seen that both the constants A and b in the
emission formula are increased thereby. At any given temperature the effect of an increase of A alone is to increase the

emission, to which indeed it is proportional, whilst an increase in *b* diminishes the emission. It appears, however, that the changes in the two constants are of such a magnitude as to cause in combination a reduction of the current. A large number of experiments, under the most varied conditions, have been made by Langmuir, who has found no exception to the rule that the *saturation* current from tungsten in presence of small amounts of any gas is never greater than the *saturation* current under the best vacuum conditions at the same temperature. This result is of the greatest importance. The only gases which have been found not to affect the values of the saturation currents are the inert gases of the argon group, mercury vapour and hydrogen, when pure. All the other gases tried have been found to reduce the value of the saturation current. The effect of all these gases on tungsten is thus the exact opposite of that of hydrogen on platinum. Under some conditions the chemically active gases may appear to increase the emission from tungsten, but this is a spurious effect due to the fact that true saturation has not been attained. Under such conditions the positive ions liberated by impact ionization in the gas may admit of a nearer approach, under a given potential difference, to the saturation value.

THERMIONIC CURRENTS FROM VARIOUS MATERIALS IN GASES AT HIGH TEMPERATURES.

Interesting experiments dealing with a number of substances have been made by Harker and Kaye.[1] They examined the conductivity between two cylindrical electrodes inside a carbon tube at temperatures between 1400° C. and 3000° C. At the higher temperatures the conductivity is very great, the currents being proportional to the voltage up to 10 volts potential difference, and increasing rapidly with temperature. When one of the electrodes is kept cold, and there is no applied potential difference between them, there is, in general, a considerable discharge of negative electricity in the direction from the hot to the cold electrode. With fresh electrodes at the

[1] " Roy. Soc. Proc., A.," Vol. LXXXVI, p. 379 (1912); Vol. LXXXVIII, p. 344 (1913).

lower temperatures the direction of this discharge is reversed and corresponds to a positive emission from the hot electrode. The negative effect at higher temperatures is greater on first heating, presumably owing to the presence of volatile impurities which are more efficient in this respect. The smaller steady currents finally obtained were about 0·15 ampere at the highest temperatures. These experiments were made at atmospheric pressure, and the currents were found to be much the same in an atmosphere of nitrogen, hydrogen, or furnace gases. The electromotive force between the hot and the cold electrode was found to be 1·8 volts.

In the second paper the authors investigate the emission from strips of platinum, iridium, iron, tantalum, nickel, copper, brass, and carbon at temperatures up to the melting-points of the metals in an atmosphere of nitrogen at pressures from 1 mm. to atmospheric. The strips were heated by an alternating current, and no external potential difference, other than that arising from the alternating circuit, was applied to drive the thermionic currents to the surrounding cylindrical electrode. The current from platinum under these conditions diminished with rising pressure. As a rule small positive emissions were observed at low temperatures. These became negative at high temperatures, and varied with the temperature in the same way as such currents have been found to do in general. With nickel, copper, and brass positive emissions only were detected.

Kaye and Higgins [1] have examined the currents, in an atmosphere of nitrogen at atmospheric pressure, which flow from a carbon crucible containing various substances to the walls of a surrounding carbon-tube furnace. Simultaneously they measured the conductivity of the furnace vapours present by means of an auxiliary electrode. The substances tested include: baryta, lime, soda-lime, strontia, magnesia, silica, alumina, ferric oxide, tin, aluminium, iron, copper, and brass. The temperatures varied from 2000° C. to 2500° C. Brass gave a large positive emission. All the other substances increased the

[1] "Roy. Soc. Proc., A.," Vol. XC, p. 430 (1914).

negative emission above the value proper to the carbon crucible. The observed currents varied from 0·1 to 1·0 ampere.

The conditions in most of the experiments just described
are so complicated that it is difficult to disentangle the various
causes which might give rise to the observed effects. No
doubt these are partly caused by electrons and ions emitted by
the hot surfaces on account of the high temperature, but they
may also be due partly to ions emitted by chemical action.
The actions are also greatly complicated by the presence of
the hot vapours which must have properties similar to those
of flames.

THE RELATIVE IMPORTANCE OF VARIOUS FACTORS IN CAUSING THE EMISSION OF ELECTRONS FROM HOT BODIES.

We have already considered three possible causes of electronic emission, namely: the escape of the electrons owing to
the purely thermal increase of their kinetic energy (p. 27), the
liberation of electrons as one of the products of chemical action
(p. 56), and the complete photoelectric emission (p. 110). We
have seen (p. 115) that the available photoelectric data indicate that the last of these is too small to account for the
emissions which have been observed from hot bodies; so that
unless and until fresh observations are made which tend to
conflict with this conclusion, it does not appear necessary to
consider this particular question further.

It remains to deal with the relative claims of the purely
thermal effect and of chemical action. Some years ago a
number of writers advocated the view that all the observed
effects were attributable to chemical action. The case for this
position is, briefly, as follows :—

We have seen in Chapter II that any electronic emission
arising from chemical action would be likely to follow a law of
temperature variation practically identical with that required
by the purely thermal effect; so that the fact that the theoretical law is satisfied by the experimental results offers no
criterion for distinguishing between the two views. In certain
cases there is some evidence that electrons are liberated as a

direct result of chemical action between solids and gases. The experiments of Haber and Just[1] have shown that when the alkali metals, their alloys, or amalgams, are attacked by oxygen, hydrochloric acid gas, phosgene gas, water vapour, and certain other chemically active gases or vapours, electrons are liberated in considerable quantity. It may be urged (cf. Chapter IX, p. 307) that in reality this also is a thermal emission, caused by a local increase of temperature in the surface layer arising from the heating caused by the chemical action.   Most of Haber and Just's experiments were made with drops of various amalgams and with the liquid alloy of sodium and potassium. From a determination of the amount of chemical action occurring they calculate that in a particular experiment the heat generated by the chemical action was not sufficient to raise the temperature of the whole of a drop more than 2° C. But it is clear that the temperature of the surface layers must have been raised to a very much greater extent, and as it is only the temperature of the surface layer which is of any account if the effect is a purely thermal one, it cannot be said that the experiments so far made by these authors prove that the emission is a direct consequence of chemical action. Similar conclusions to those of Haber and Just have been reached by Fredenhagen,[2] who found that the currents he observed when the alkali metals were heated could be reduced to very much smaller values by the careful elimination of gases. The smallest currents recorded by Fredenhagen are, however, not smaller than those calculated[3] by an application of the considerations on p. 42 to the known emission from platinum or tungsten, in spite of the strong electropositive character of the alkali metals. The large values are restored, at least partially, when small quantities of gases, and especially of oxygen, are allowed to come in contact with the metal. But when one remembers the extraordinary sensitiveness of the emission to changes of temperature and that, in any event, the effect is a purely superficial one, it is questionable whether the observed enhance-

[1] "Ann. der Physik," Vol. XXXVI, p. 308 (1911).
[2] "Verh. der Deutsch. Physik. Ges.," Jahrg. 14, p. 386 (1912).
[3] Cf. O. W. Richardson, "Phil. Mag.," Vol. XXIV, p. 742 (1912).

ment of the emission may not be due to the local increase of temperature caused indirectly by the chemical action. It is, of course, abundantly proved that an emission is caused by chemical action in these cases, but it is extremely difficult to be sure that the effect is the direct result of the chemical action and is not caused indirectly by the heat generated at the surface. (See, however, p. 313.) In any event in Fredenhagen's experiments his currents were not saturated, and the reduction in the currents which he observed with improving vacua may well have been caused by the charging up of the glass walls of his apparatus and by space charge effects.

The remaining cases which have been cited as examples of the emission of electrons by chemical action are the oxidation of calcium and the emission from incandescent carbon. Fredenhagen[1] has put forward the view that the activity of the lime-covered cathode is caused by the recombination of the calcium and oxygen which are separated by electrolysis during the passage of the current. The arguments in favour of this view have already been dealt with on p. 98 where they were not found to resist a critical examination successfully. We have already seen (p. 100) that Horton, who made a direct test of the question, was unable to detect any emission from calcium arising directly from oxidation. Wehnelt,[2] who has devoted much attention to the lime-covered cathode, has recently expressed the opinion that in the case of this material there is no evidence which would favour a chemical rather than a purely thermal cause for the origin of the emission. Germershausen[3] has shown that the emission from lime is increased by the removal of every trace of gas from its surroundings. Under these conditions the discharge becomes very similar to that from tungsten as observed by the writer and by Langmuir. The experiments made in the Research Laboratory of the American Telephone and Western Electric Companies described on p. 101 would appear finally to dispose

[1] " Leipziger Ber.," Vol. LXV, p. 42 (1913).
[2] " Physik. Zeits.," Jahrg. 15, p. 558 (1914).
[3] *Ibid.*, Jahrg. 16, p. 104 (1915).

of any such secondary hypothesis as to the origin of the emission from oxy-cathodes.

The emission from carbon has been attributed to chemical action between the carbon and traces of gaseous contamination by Pring and Parker,[1] and by Pring.[2] Using comparatively large rods of carefully purified carbon they found that the negative discharge to a small electrode in the neighbourhood of the rod diminished progressively as the gases were removed from the rod by continuous heating. The currents finally obtained at the highest temperatures were very much smaller than those recorded by other observers with carbon (see pp. 81 and 88). An application[3] of the considerations developed on p. 75, however, shows that the magnitude of the heating currents and the geometrical arrangement of the apparatus used by these authors were such that no electrons at all would be able to reach the electrode at the higher temperatures. Thus there is no difficulty in accounting for the smallness of the observed currents on the purely thermal theory of the emission. The only difficulty, which is present on any theory of the origin of the electrons, is to explain why the observed currents were not actually zero. The small currents can be accounted for[4] if it is supposed either that traces of gas present interfere with the motion of the electrons or that some of the electrons combine with uncharged molecules or atoms of the gas to form negative ions whose motion is almost unaffected by the magnetic field. Either of these assumptions would explain the fact that the observed currents are increased by the admission of traces of various gases. The relatively large effects produced by very small amounts of gas are in favour of the second hypothesis, which also explains the relative efficiency in this respect of the various gases tested. Another factor which would tend to make the currents observed in these experiments too small is the effect of the mutual repulsion of the electrons. Moreover, the results

[1] " Phil. Mag.," Vol. XXIII, p. 192 (1912).
[2] " Roy. Soc. Proc., A.," Vol. LXXXIX, p. 344 (1913).
[3] O. W. Richardson, *ibid.*, Vol. XC, p. 174 (1914).
[4] Loc. cit.

of the experiments are in complete disagreement with the results of experiments made with well-glowed-out carbon filaments by Deininger[1] in 1908, and of the more recent experiments of the writer and of Langmuir. In the last two cases the precautions described on p. 14 were taken in preparing the bulbs, and although it is not claimed that every trace of gas was got out of this very difficult substance, the conditions were much better in this respect than in the experiments of Pring and of Pring and Parker. The same claim can almost certainly be made for Deininger's work. In fact, it is quite impossible to attain good vacuum conditions with the large quantities of hot material and the other arrangements used by Pring. Taking all the facts into consideration they appear to the writer to afford no support to the contention that the emission from carbon has anything whatever to do with chemical action. It may be that such a chemical effect exists, but its existence is not demonstrable, or even rendered probable, by the evidence which has been submitted.

Thus there is no case in which it has been established with certainty that chemical action is the direct and immediate cause of an emission of *electrons*. The majority of chemical actions between solids and gases certainly do not give rise to electrical effects of this kind to any appreciable extent (see Chap. IX). The only case in which the evidence renders the occurrence of electron emission as a chemical effect probable is that of the alkali metals. The experiments of Haber and Just do, on the whole, indicate a balance of probability in favour of a direct chemical effect in this case, although, in the judgment of the writer, they cannot be held to establish it with certainty.

The advocates of the chemical point of view have held that the emissions usually observed are due to actions between the hot metal and minute traces of residual gas and not to chemical actions on any considerable scale. In support of this it may be urged that, as the effect is a purely superficial one, a small quantity of gas will exert as large an effect as a

[1] "Ann. der Physik," Vol. XXV, p. 285 (1908).

greater amount, down to a certain limit. On the other hand, it cannot be said that a comparison of the specific effects of different gases lends any support to the chemical theories. In the case of platinum, the only gas which causes any considerable increase in the emission is hydrogen, and although there is probably some chemical combination in this case it certainly is not of a violent character. In the case of tungsten all gases which act chemically on the metal have been found to reduce the emission and not to increase it.

The difficulty, discussed on pp. 83 *et seq.*, of determining the precise values of the constants A and *b*, and the dependence generally of the emission on factors which are difficult to control and to specify, has been held to favour the view that these effects are caused by the interaction between the hot bodies and traces of gaseous contamination of uncertain composition. It would, however, seem more reasonable to attribute these features of the phenomenon to the fact that it is of an entirely superficial character and is very sensitive to changes in the nature or composition of the surfaces. For example, the admission of oxygen will coat the hot metal with a layer of oxide. If, as appears to be the case with calcium, the oxide is more active thermionically than the metal, the emission in presence of oxygen will exceed the normal value. If the oxide is inactive its presence will tend to prevent the electrons escaping from the metal and will thus reduce the emission. The effect of oxygen on tungsten is probably of this nature. Minute traces of gas would be sufficient to produce effects of this kind, and if the composition of the gas were uncertain and variable, the effects would be correspondingly so. A similar difficulty arises in other superficial phenomena, such as the photoelectric effect, surface tension, and optical reflexion, although, as a rule, it is not so marked. This is on account of the extreme sensitiveness of the thermionic emission to small changes in the work required for an electron to escape. There is, in fact, no comprehensive body of evidence supporting the view that interaction with gases is an invariable and direct cause of thermionic emission; the evidence that gases act indirectly by modifying the quantity of the emission quite generally is of a much stronger character. In this connexion the close relationship between electron emission and contact potential difference, which is required on theoretical grounds, and the sensitiveness of both these phenomena to superficial contamination, should also be kept in mind.

Experiments made by the writer [1] have shown that the emission from tungsten in a good vacuum is a property of the element itself, and cannot be attributed to chemical or other secondary actions between the tungsten and traces of other contaminating material. The advantages of tungsten in investigations of this character have been alluded to already (p. 132). The tests were made with experimental tungsten lamps carrying a vertical filament of ductile tungsten which passed axially down a concentric cylindrical electrode of copper gauze or foil. The tungsten filaments were welded electrically in a hydrogen atmosphere to stout metal leads. These in turn were silver-soldered to platinum wires sealed into the glass container. The lead to the copper electrodes was sealed into the glass in the same way. The lamps were exhausted with a Gaede pump for several hours. During this time they were maintained at 550° to 570° C. by means of the vacuum furnace described in Chapter I. The duration of this exhaustion varied from 8 to 24 hours with different bulbs. It was continued until the apparent evolution of gas was very small and practically constant. This small final development of gas, which appeared to persist indefinitely, is believed to be due to the dissociation of the glass walls of the tube and to the diffusion through them of gases from the vacuum furnace which could only be exhausted to about 1 cm. pressure. The exhaustion was completed by means of liquid air and charcoal, the tungsten filament meanwhile being glowed out by means of an electric current at over 2200° C. Most of the tests were made after the furnace had been opened up and the lamps allowed to cool off. This treatment has been found completely to stop the emission, under the relatively slight heating caused

---

[1] " Phil. Mag.," Vol. XXVI, p. 345 (1913).

by the radiation from the hot filament, of gases from the walls of the tubes and from the cold electrodes which had previously formed such a persistent source of difficulty in experiments with hot wires.

Although the filaments used were quite thin (about 0·007 cm. diameter), these lamps were found capable of being run so as to give thermionic currents of about 0·1 ampere for hours. Tests were made covering the following alternative possible causes of the emission :—

1. That the emission is caused by the evolution of gas from the filaments.

In one experiment the tube was shut off by a mercury trap and the gases allowed to accumulate. The filament gave an electronic current of 0·50 ampere continuously for 30 minutes. The pressure of the gas which had accumulated was less than 10^{-7} mm. and was too small to measure. Taking into account the volume of the bulb, these figures show that for every molecule of gas evolved $2·6 \times 10^8$ electrons were emitted. No conceivable process could cause so many electrons to arise from each gas molecule.

2. That the emission is caused by chemical action or some other cause depending on impacts between the gas molecules and the filaments.

If for purposes of computation we consider the gas to be hydrogen, which is the most unfavourable assumption, since this gas makes most collisions, the data of the last experiment show that 15,000 electrons would have to arise every time a molecule impinged on the filament. This number is of course quite prohibitive. Moreover, in certain other experiments quite appreciable changes in the gas pressure caused no change in the emission.

3. That the emission is a result of some process involving consumption of the tungsten.

In these experiments there is a loss of tungsten from the filament which is believed to be due to evaporation. The loss was determined by measuring the change in the resistance of the filament. At the same time the thermionic current was measured, giving the number of electrons emitted. In one experiment it was found that for each atom of tungsten lost 984,000 electrons were emitted. In this case the mass of the electrons emitted was *three times* the mass of tungsten lost. This experiment and others similar to it show conclusively that the emitted electrons must have flowed into the tungsten from outside points of the circuit.

4. That the emission is caused by interaction with some condensible vapour which does not affect the McLeod gauge.

This explanation is cut out by the fact that the currents are not affected when the tube is cut off from the liquid air and charcoal and the hypothetical vapours allowed to accumulate.

These experiments have not been accepted as conclusive by Fredenhagen[1] and by Horton[2] on the ground that they still leave open the possibility that the emission is due to interaction with the tungsten of some substance present in the filaments. In regard to this suggestion it is to be remembered that the assumption of the presence of foreign substances in the filament is a pure hypothesis. It is very unlikely that any gaseous substance, and most substances are gaseous under these conditions, could remain in a thin filament kept at over 2200° C. in a vacuum of 10^{-6} mm. pressure for a long time. The behaviour of the filaments during the experiments is distinctly opposed to this suggestion. When they are first glowed out there is a considerable evolution of gas lasting for a few seconds, and after that nothing. When the filaments are sealed in a small closed tube and allowed to disintegrate through overrunning no gas is evolved. There is good evidence that the small quantities of gas which occasionally appear in experiments of this kind come from the walls of the tube and the relatively heavy parts of the metal electrodes owing to inadequate preliminary treatment. The only impurities which would seem to have any chance of remaining in the filaments during these experiments are the highly refractory elements such as molybdenum, tantalum, carbon, thorium, etc. Even these would be expected gradually to disappear, and *there is no evidence of any progressive change in the emission at constant*

[1] " Phys. Zeits.," Jahrg. 15, p. 19 (1914).
[2] " Phil. Trans., A.," Vol. CCXIV, p. 278 (1914).

temperature with properly prepared tubes. In any event it is questionable whether their presence would help the chemical theory, which would then be reduced to the position of admitting the existence of an emission from alloys but not from the pure metals. That the emission cannot be attributed to the commoner gases is also shown conclusively by the experiments (see p. 141) of Langmuir, who found that they all reduced the emission, except the inert gases which left it unaltered.

These experiments with tungsten definitely exclude chemical action as the cause of the emission from this substance. Such, at least, is the considered judgment of the writer. Although equally searching tests have not been made with other materials, a general survey of the phenomena does not indicate any definite connexion with chemical action, certainly in the case of the refractory elements. This is supported by the results of Langmuir,[1] who finds that with tantalum, molybdenum, carbon, and platinum as well as tungsten, the emission is increased with progressive elimination of gaseous contamination and corresponding freedom from liability to chemical action. There is, of course, no compelling reason to expect a purely thermal origin for the effects in all cases. It may be that in the case of the alkali metals such effects as have been observed are due entirely or chiefly to chemical action; but this has not yet been proved, certainly not with anything like the thoroughness of proof of the contrary proposition in the case of tungsten. If the emission is ever caused by chemical action we should expect this type of effect to be exhibited by the alkali metals, where the reactions are much more vigorous than with the refractory elements, as is shown by the very much greater heat liberation per gram equivalent.

There is another argument, to which great weight should be attached, which is definitely against a chemical origin of the effects exhibited by the refractory elements. We have seen that the variation of the emission with temperature enables us to form an estimate of the energy change associated

[1] "Phys. Rev.," Vol. II, p. 484 (1913).

with the liberation of one electron. We shall see in the next chapter that more direct methods are available for determining this quantity, both from the absorption of heat when electrons are emitted, and from the liberation of heat when electrons are absorbed. All three methods give consistent results, and show that the quantity in question is very considerable.

If we compare this heat change per gram equivalent of electrons, with the heat liberated per gram equivalent in various chemical reactions, we find, in the case of tungsten or platinum, that it is about equal to the corresponding quantity for the most vigorous chemical actions known, such as the combination of the alkali metals with the haloids, and is far greater than the heat of any known reaction of the elements under consideration. Thus the rate of variation of the emission with temperature is right for the physical theory of the phenomena, but is wrong, so far as we can judge, for the chemical theory. It is desirable at present to restrict this argument to the more refractory elements which are less active chemically, as the thermionic data for the more electropositive elements cannot be considered to be known with sufficient definiteness. However, such data as are available for the more electropositive metals all go to show that this quantity is less for them than for chemically inert metals like platinum which is obviously a very difficult matter to explain on any chemical theory.

CHAPTER V.

ENERGETICS OF ELECTRON EMISSION.

1. THE KINETIC ENERGY OF THE EMITTED ELECTRONS.[1]

WE saw in Chapter II that the law of temperature variation of the emission of electrons could be deduced in various ways from a consideration of the properties of the atmosphere of electrons in equilibrium with hot bodies present in a vacuous enclosure. The essential and important results of these theories have been very fully confirmed by the experimental results already described. The further consideration of such atmospheres of electrons suggests certain other important properties of the streams of emitted electrons which have not yet been discussed. We have seen that the electron atmospheres are in all respects analogous to a gas, the only important differences arising from the much smaller value of the molecular weight, and, owing to the fact that the electrons carry an electric charge, the much greater value of the intermolecular forces. Just as in the case of gases the modification of the pressure due to the intermolecular forces becomes negligible at very low pressures, we see that the pressures due to very attenuated electron atmospheres will be the same as those which would be exerted if the electrons were uncharged. In point of fact, the electron concentrations to be dealt with are excessively small; so that the pressures will be given by the law of a perfect gas

$$p = nkT, \qquad . \qquad . \qquad . \qquad . \qquad (1)$$

as we have already assumed. In (1) n is the number of electrons per c.c. in the atmosphere in equilibrium, and k is Boltzmann's constant.

---

[1] In the light of recent experiments some of the statements in this and the following section may need qualification. Cf. p. 172.

We know also, from the principles of the dynamical theory of gases, that in such an atmosphere the average kinetic energy of each molecule is proportional to the absolute temperature and equal to $\frac{3}{2}kT$, and that the velocities of the different molecules are distributed amongst them in accordance with Maxwell's Law. The same conclusions will apply to the streams of electrons as they are emitted from a metal surface, even when they are allowed constantly to flow away, and there is no possibility of the attainment of steady equilibrium conditions. This follows, since any such change as that contemplated will not affect the conditions which determine the emission of the electrons. The emitted stream will thus have the same properties whether the external conditions are those of equilibrium or not. When the conditions are those of equilibrium it follows, from the principles of the dynamical theory of gases, that the emitted stream must have the properties specified above; whence it follows that this statement as to the properties of the emitted stream must be true in general.

This conclusion is valid even if the principles of the dynamical theory of gases are not universally applicable, for instance, if the emission of the electrons is governed by the principles of the quantum theory in some such manner as they are developed on p. 37; for the equilibrium concentration of the external electrons in these cases is so small that the principles of the classical dynamics will still apply to them even if the phenomena as a whole are governed by the quantum theory. On the other hand, if the distribution of velocity amongst the emitted electrons is governed by Maxwell's Law, it does not follow that the same thing is true of the distribution of velocity amongst the free electrons inside the hot body, for the concentration of these must be of an entirely different, and in all probability much higher, order of magnitude.

The result of this argument may be summarized as follows: We expect, as a consequence of the theories developed in Chapter II, that the distribution of kinetic energy amongst the electrons in the emitted stream will be identical with that amongst those molecules of a gas, at the same temperature as the hot body, which leave either side of any surface in the gas in any definite interval of time. In accordance with Maxwell's Law the average energy of the electrons emitted is $2k\mathrm{T}$, and, if the emitting surface is taken perpendicular to the axis of x and u, v, w are the velocity components of an electron parallel to x, y, and z respectively, then the number emitted in unit time with velocity components between u and $u + du$ is

$$\mathrm{N}_u du = \mathrm{N} \,.\, 2hmue^{-hmu^2}du, \qquad . \qquad . \quad (2)$$

the number with velocity components between v and $v + dv$,

$$\mathrm{N}_v dv = \mathrm{N} \,.\, \sqrt{\frac{hm}{\pi}}\, e^{-hmv^2}dv, \qquad . \qquad . \quad (3)$$

and the number with velocity components between w and $w + dw$,

$$\mathrm{N}_w dw = \mathrm{N} \sqrt{\frac{hm}{\pi}}\, e^{-hmw^2}dw, \qquad . \qquad . \quad (4)$$

where N is the total number emitted in unit time, m is the mass of an electron, and $h = (2k\mathrm{T})^{-1}$. It will be noticed that the average kinetic energy of the emitted electrons is $2k\mathrm{T}$ and not $\frac{3}{2}k\mathrm{T}$, the average kinetic energy of the electrons (or molecules) in unit volume in equilibrium. The larger value arises from the fact that the more rapidly moving particles occur more frequently in an emitted stream than in the number present in a volume selected at random under equilibrium conditions.[1]

These conclusions have been tested in a large number of experiments made by the writer, partly in collaboration with F. C. Brown. The first investigation, made by Richardson and Brown,[2] is concerned only with the component of velocity u normal to the emitting surface. The apparatus used is shown in section in Fig. 15.

The emitting surface was that of a small piece of thin platinum foil H heated electrically. The foil nearly filled a small hole at the centre of the metal plate L, the upper surfaces of L and H being flush with one another. The heating

[1] Cf. O. W. Richardson, "Phil. Trans., A.," Vol. CCI, p. 502 (1903); "Phil. Mag.," Vol. XVIII, p. 695 (1909).

[2] *Ibid.*, Vol. XVI, p. 353 (1908).

current was let in through t_1, t_2, which were connected by a high resistance shunt not shown in the figure. The shunt was provided with a sliding contact which could be connected through the metal base B to L. In this way the middle of the strip H could be kept at the same potential as the surrounding plate L. This device, for controlling the potential of inaccessible parts of an enclosed apparatus, carrying an

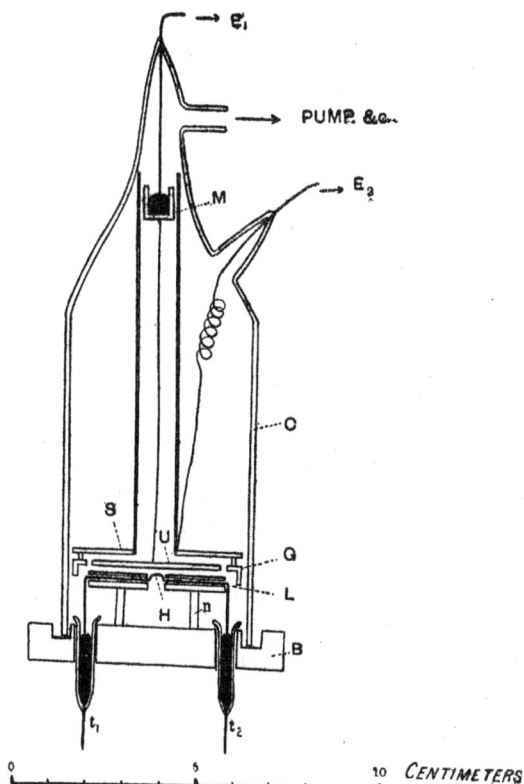

FIG. 15.

electric current, is often useful in experiments of this character. Opposite L is a parallel plate U covered with platinum, to avoid effects arising from contact difference of potential, and provided with a guard ring G and electrostatic shield S. U is connected to the insulated quadrants of a sensitive electrometer, whose time rate of deflexion measured the number of electrons passing from H to U. The temperature of H was controlled, and estimated, by measuring its resistance in the manner described in Chapter I, p. 16. The electron currents from H to U were measured when different potentials were applied so as to *oppose* their passage.

Now let us consider the theory of this experiment supposing, first of all, that the planes U and L are infinite in extent. The plates are maintained at fixed potentials; so that the electric intensity is everywhere normal to them, i.e. parallel to the x axis, and constant. If V is the potential at any point x, y, z between the two plates the equations of motion of an electron at that point are—

$$m\frac{\partial^2 x}{\partial t^2} = m\frac{\partial u}{\partial t} = -\epsilon\frac{\partial V}{\partial x}, \quad . \quad . \quad . \quad (5)$$

$$m\frac{\partial^2 y}{\partial t^2} = m\frac{\partial v}{\partial t} = 0 \text{ and } m\frac{\partial^2 z}{\partial t^2} = m\frac{\partial w}{\partial t} = 0 \quad . \quad (6)$$

From (6), the v and w velocity components are constant during the motion, and using the factor $u = \frac{\partial x}{\partial t}$ to integrate (5),

$$u^2 = u_0^2 - \frac{2}{m}\epsilon V, \quad . \quad . \quad . \quad . \quad (7)$$

if u_0 is the emission value of u at the lower plate, where V = 0. If the upper plate is charged negatively, so that the potential difference tends to oppose the passage of the electrons, both ϵ and V are negative; so that the product ϵV is positive. u will be reduced to zero on reaching a point at which $V = mu_0^2/2\epsilon$; after passing this point the electron will return to the lower plate. If V_1 is the difference of potential between the plates, we see from (7) that an electron will get as far as the upper plate, provided

$$u_0^2 \geqq \frac{2}{m}\epsilon V_1, \quad . \quad . \quad . \quad . \quad (8)$$

otherwise it will return to the lower plate. Thus with an opposing difference of potential equal to V_1, only those electrons will contribute to the current from the upper plate which satisfy (8). It follows that if $F(u_0)du_0$ is the proportion of electrons emitted for which the u component of velocity lies between u_0 and $u_0 + du_0$, $f(v_0)dv_0$ and $f(w_0)dw_0$ denoting the corresponding functions for the v and w components, the current from the upper plate will be given by

$$i = C\frac{\partial V_1}{\partial t} = N\epsilon \int_{\sqrt{2\frac{e}{m}V_1}}^{\infty} F(u_0)du_0 \int_{-\infty}^{\infty} f(v_0)dv_0 \int_{-\infty}^{\infty} f(w_0)dw_0, \quad (9)$$

where C is the capacity of the electrometer and its connexions and N is the number of electrons emitted, with any velocity, in unit time. If the upper plate has a finite radius r we have to take account of the fact that the radial velocity may be sufficient to take some of the electrons a horizontal distance greater than r before the vertical distance x_1 between the plates has been covered. Under these circumstances equation (9) is altered to

$$i = N\epsilon \int_{\sqrt{2\frac{e}{m}V_1}}^{\infty} F(u_0)du_0 \int_0^{\frac{1}{2}\frac{r}{x_1}\left(u_0 + \sqrt{u_0^2 - 2\frac{e}{m}V}\right)} F(W)dW, \quad (10)$$

where $F(W)dW$ denotes the probability of the radial velocity $W = \sqrt{v_0^2 + w_0^2}$ lying between W and W + dW. However, the difference between (9) and (10) was negligible in the experiments referred to; so that we need consider only the simpler expression (9).

If Maxwell's Law holds, the values of $NF(u)_0\ du_0$, etc., will be given by the right-hand sides of the corresponding equations (2) to (4) and by substituting these values in (9)

$$i = N\epsilon e^{-2hV_1\epsilon} = i_0 e^{-2hV_1\epsilon} \ . \qquad . \qquad . \quad (11)$$

if i_0 is the value of i when $V_1 = 0$. Remembering that $h = (2kT)^{-1}$, and taking logarithms, we obtain

$$\log i/i_0 = -\frac{V_1\epsilon}{kT} = -\frac{v\epsilon}{RT}V_1 \qquad . \qquad . \quad (12)$$

where v is the number of molecules in 1 c.c. of a perfect gas at $0°$ C. and 760 mms. pressure, and R is the constant in the equation $pv = RT$ calculated for this quantity of gas. We have seen that both $v\epsilon$ and R are well-known physical constants, being equal to 0·4327 e.m. unit and 3·711 × 10⁸ erg./deg. C. respectively.

The results of one of the experiments are plotted in Fig. 16. The points shown thus : ⊙ give the current i as ordinates and the points shown thus: × the values of log i. The abscissæ are the values of the corresponding opposing potentials in each case. From (12) we see that log i should be a linear function of V_1 at constant temperature. This re-

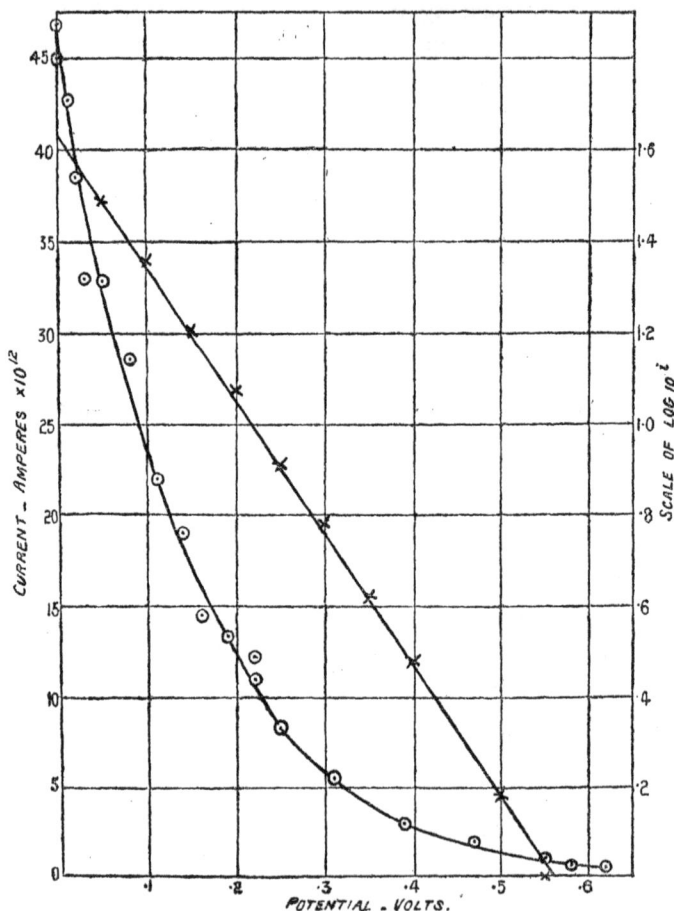

FIG. 16.

quirement is ·satisfied very accurately by the points: × on the diagram. From the slope of the straight line, knowing T and assuming that $\nu\epsilon = \cdot 4327$, we can calculate the value of the constant R. The values of R obtained in this way under a variety of conditions are collected in the following table.

The numbers in the last column are a little higher than' those given in the original paper, where an inaccurate value of ν_t was used.

Treatment of Platinum before Heating.	Pressure mms.	Absolute Temperature (°K.).	Maximum Current (amperes).	R. ergs./°C.
16 hours' heating . .	0·015	1556	4·7 × 10⁻¹¹	4·36 × 10³
	0·008 0·009 }	1473	1·2 × 10⁻¹¹	4·46 × 10³
Just after lime was placed on the platinum . }	0·006 0·06 }	1503	3 × 10⁻¹¹	3·72 × 10³
Just after hydrogen was let into the vacuum . .	0·04	1553	4 × 10⁻¹¹	3·83 × 10³
About 35 hours' heating .	0·015	1660	1·4 × 10⁻¹¹	3·08 × 10³
About 30 hours' heating .	0·01	1560 ·	3 × 10⁻¹²	3·29 × 10³
Highly charged with negative electricity and strongly heated subsequently . .	0·02	1840	4 × 10⁻¹¹	3·40 × 10³
Highly charged with positive electricity and strongly heated subsequently .		1813	1 × 10⁻¹¹	3·61 × 10³

The mean of the numbers in the last column gives R = $3\cdot719 \times 10^3$ as compared with the theoretical value $3\cdot711 \times 10^3$. No doubt this excellent agreement is partly accidental, but it shows that the average kinetic energy of the emitted electrons is close to that of the molecules of a gas at the same temperature as the hot body. The fact that the linear relation be-tween log i and V_1 is satisfied shows not only that the average energy is the same, but also that the energy is actually distributed among the electrons in exactly the same way as it would be distributed among the molecules of a monatomic gas at the same temperature. The experiments considered above have only proved these statements to be true so far as the part of the kinetic energy is concerned which depends on the component of velocity perpendicular to the emitting surface.

Richardson and Brown also made a few observations on certain other substances with the object of ascertaining the law of distribution of velocity among the electrons emitted by them, using the same or a similar method. The substances tested were : platinum saturated with hydrogen so as to give a large emission, platinum coated with lime, and the liquid alloy of sodium and potassium. These experiments were not at all satisfactory, but, so far as they went, they indicated that the distribution of energy amongst the electrons emitted by these bodies was not in accordance with Maxwell's Law.

There are a number of ways in which such a result might arise exceptionally without vitiating any general principle; but it is not worth while to discuss the matter further in the absence of more satisfactory experimental evidence. The importance of the subject makes it very desirable that more experimental work should be done with these substances.

The distribution of velocity for the components *parallel* to the emitting surface was examined by the writer[1] using a different type of apparatus. A vertical section of one form of this is shown in Fig. 17. The parallel metal plates A, B are provided with narrow central parallel slits perpendicular to the plane of the figure. A narrow platinum strip, provided with an arrangement for keeping it flat, worked from outside the apparatus, almost fills the slit D. The platinum is heated electrically and its front surface is flush with that of the plates.

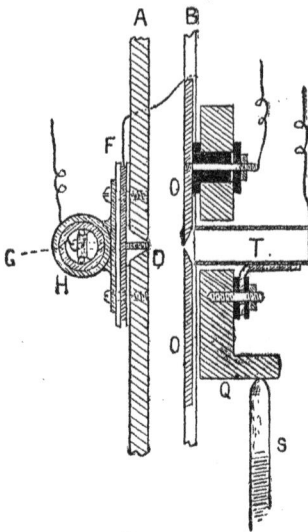

FIG. 17.

The electrons emitted by the strip are carried by the electric field to the opposite plate OO, but some of them pass through the slit into the box-shaped electrode T, which is insulated from the plates. All the parts OTQO are rigidly bolted together and can be moved up and down through known distances by means of the accurate screw S. By means of suitable electrometer connexions (see p. 213) the number of electrons passing through the slit and the number reaching the plates can be measured simultaneously. These quantities were measured for different vertical displacements of the slit in OO relative to the level of D. This information enables the distribution of the vertical component of velocity of the emitted electrons to be ascertained.

[1] " Phil. Mag.," Vol. XVI, p. 890 (1908); Vol. XVIII, p. 681 (1909).

In these experiments an accelerating difference of potential V_1 is applied between the plates A and B so as to pull the electrons from D towards OO. Since it can be shown that the effect of the mutual repulsion of the electrons is negligible with the small currents used, it follows that if the electrons were emitted with no velocity component parallel to the surface of emission they would travel in straight lines normal to the plates. Under these conditions the graph of y, the ratio of the current through the slit to the total current received by both the slit and the plates, against x, the vertical displace-

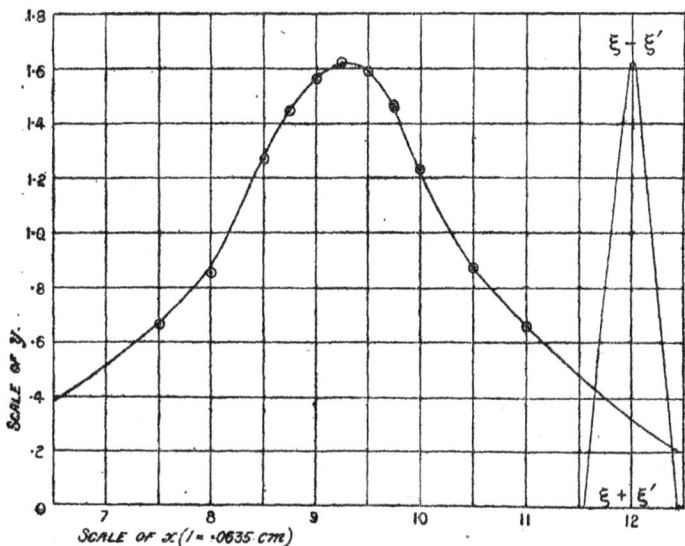

FIG. 18.

ment of the slit, would consist of three inclined straight lines as is shown on the right-hand side of Fig. 18. The observed graph is that drawn through the points marked thus : ⊙ and shows very clearly the spreading out of the electrons owing to the vertical component of velocity.

Now consider the case when the electrons are emitted with initial velocity components u_0 v_0 w_0. Let us take the planes to be perpendicular to the axis of z and w, and the axis of x and u to be parallel to the vertical line in Fig. 17. If Z is the electric intensity arising from the difference of potential V_1

between the plates, the equations of motion of an electron are

$$m\frac{\partial^2 x}{\partial t^2} = 0 \quad m\frac{\partial^2 y}{\partial t^2} = 0 \quad . \quad . \quad . \quad (13)$$

and

$$m\frac{\partial^2 z}{\partial t^2} = Z\epsilon = -\epsilon\frac{\partial V_1}{\partial z} \quad . \quad . \quad (14)$$

If $t = 0$ when the electron starts from the strip, the initial conditions (at $t = 0$) are

$$\frac{dx}{dt} = u_0, \frac{dy}{dt} = v_0, \frac{dz}{dt} = w_0, \ x = x_0, y = y_0, \text{ and } z = z_0.$$

For the present problem we are concerned only with the x and z displacements, i.e. with the motion projected into the plane of Fig. 17. Integrating equation (14) and the first of equations (13) subject to the initial conditions above, and eliminating t, we get

$$u_0 = \tfrac{1}{2}w_0\frac{x_1 - x_0}{z_1}\left[1 + \left(1 + \frac{V_1\epsilon}{\tfrac{1}{2}mw_0^2} \right)^{1/2} \right], \quad . \quad (15)$$

where z_1 is the perpendicular distance between the planes, and x_1 is the vertical level at which an electron emitted at the level x_0 with velocity components u_0 w_0 strikes the opposite plane. Considering electrons setting out with different values of u_0, those for which u_0 exceeds the right-hand side of (15) will strike the plane at a level higher than x_1, and those with smaller values of u_0 at a lower level than x_1. It follows from this that if ξ' is the width of the hot strip and ξ that of the slit, both supposed to be of indefinite length, the current passing through the slit at the level x_1 is

$$i = \frac{N_1\epsilon}{\xi'}\int_{-\tfrac{1}{2}\xi'}^{+\tfrac{1}{2}\xi'} dx_0 \int_0^\infty F(w_0)dw_0 \int_{\tfrac{1}{2}w_0\frac{x_1 - x_0 - \xi/2}{z_1}\left[1+\left(1+\frac{2V_1\epsilon}{mw_0^2}\right)^{1/2}\right]}^{w_0\frac{x_1 - x_0 + \xi/2}{z_1}\left[1+\left(1+\frac{2V_1\epsilon}{mw_0^2}\right)^{1/2}\right]} f(u_0)du_0 \quad (16)$$

where N_1 is the total number of electrons emitted by the strip, and $F(w_0)dw_0$ and $f(u_0)du_0$ are the proportions of them for which w_0 lies between w_0 and $w_0 + dw_0$, and u_0 between u_0 and $u_0 + du_0$ respectively. If ξ and ξ' are both small, as in the experiments, there are two important cases for which (16) reduces to a quite simple expression.

First suppose that $V_1\epsilon/\tfrac{1}{2}mw_0^2$ is a large quantity. It is worth while remarking that if, as we have seen is the case, Maxwell's Law holds for w_0, this condition cannot be satisfied by all the electrons; since all values of w_0 up to infinity are included in the theoretical formula. On the other hand if V_1 is of the order of 100 volts the fraction of the whole number of electrons which does not satisfy this condition is exceedingly small and may safely be neglected. When $2V_1\epsilon/mw_0^2$ is large, (16) reduces to

$$i = N_1\epsilon \, \frac{\xi}{\sqrt{\pi}} \left(\frac{hV_1\epsilon}{2z_1^2}\right)^{\frac{1}{2}} e^{-\frac{hV_1\epsilon}{2z_1^2}(x_1 - x_0)^2} \qquad . \qquad (17)$$

$$= N_1\epsilon \, \frac{\xi}{\sqrt{\pi}} \left(\frac{\nu\epsilon V_1}{4RTz_1^2}\right)^{\frac{1}{2}} e^{-\frac{\nu\epsilon V_1}{4RT}\frac{(x_1 - x_0)^2}{z_1^2}} \qquad . \qquad (18)$$

after substituting the values of $F(w_0)$ and $f(u_0)$ which are required if Maxwell's Law is to be satisfied. In (17) and (18) h, ν and R have the meanings given to them on p. 159. When the centre of the slit is opposite the centre of the strip $x_1 - x_0$, which in what follows may be denoted by the single letter x without confusion, is equal to zero; so that, if i_0 is the current through the slit when in this position,

$$i_0 = N_1\epsilon \, \frac{\xi}{\sqrt{\pi}} \left(\frac{\nu\epsilon V_1}{4RTz_1^2}\right)^{1/2} \qquad . \qquad (19)$$

Thus, dividing (18) by (19) we obtain

$$i/i_0 = e^{-\frac{\nu\epsilon V_1}{4RT}\frac{x^2}{z_1^2}} \qquad . \qquad . \qquad (20)$$

or

$$\log i = \log i_0 - \frac{\nu\epsilon V_1}{4RT}\frac{x^2}{z_1^2} \qquad . \qquad . \qquad (21)$$

Equations (17) to (21) have been shown to follow if the distribution of the vertical component of velocity at emission is distributed in accordance with Maxwell's Law. We may, however, proceed quite differently by deducing the law of distribution of velocity directly from the experimental curves. When $2V_1\epsilon/mw_0^2$ is very large it follows from (15) that the electrons which reach the opposite plane at the level x, measured from the level $x = 0$ of the narrow emitting strip, are emitted with the vertical velocity component

$$u_0 = \frac{x}{2}\left(\frac{2\epsilon_1 V}{mz_1^2}\right)^{1/2} \quad . \quad . \quad . \quad . \quad (22)$$

The part of their kinetic energy which arises from this velocity component is thus

$$\tfrac{1}{2}mu_0^2 = \epsilon V_1 x^2 / 4z_1^2 \quad . \quad . \quad . \quad (23)$$

It follows from (22) that the electrons for which u_0 lies between u_0 and $u_0 + du_0$ have a value of x which lies between $\left(\frac{2mz_1^2}{\epsilon V_1}\right)^{\frac{1}{2}}u_0$ and $\left(\frac{2mz_1^2}{\epsilon V_1}\right)^{\frac{1}{2}}(u_0 + du_0)$. Given x, to find the corresponding value of u_0, all we have to do is to multiply by a factor involving the known quantities ϵ/m, V_1, and z_1. Thus in Fig. 18 the abscissæ represent the values of u_0 as well as of x, and since the currents are proportional to the numbers of electrons, the ordinates represent the numbers corresponding to given values of u_0. Thus curves like that in Fig. 18 form a complete graphical representation of the mode of distribution of the component u_0 of velocity amongst the emitted electrons. For example, to find the average kinetic energy arising from u_0 from Fig. 18 we can proceed as follows : If y is the ordinate at any point and ν_0 the number of electrons corresponding to unit area of the diagram the number which corresponds to a strip of height y and breadth dx is $\nu_0 y dx$. The kinetic energy of these electrons is $\nu_0 y dx \times \tfrac{1}{2}mu_0^2 = \nu_0 \frac{\epsilon V_1}{4z_1^2} y x^2 dx$. The total amount of this energy is thus $\nu_0 \frac{\epsilon V_1}{4z_1^2}\int_{-\infty}^{\infty} y x^2 dx$, and the total number of electrons to which it belongs is $\nu_0 \int_{-\infty}^{\infty} y dx$. The average amount of this part of the kinetic energy pertaining to each electron is therefore

$$\frac{\epsilon V_1}{4z_1^2}\int_{-\infty}^{\infty} y x^2 dx \bigg/ \int_{-\infty}^{\infty} y dx \quad . \quad . \quad . \quad (24)$$

The two integrals may be evaluated graphically in the usual manner.

The relations (17) to (24) have been tested in various ways. The curve on the left in Fig. 18 is the curve

$$i = 1\cdot38 e^{-55\cdot1 x^2} \quad . \quad . \quad . \quad (25)$$

and is seen to pass through all the experimental points. This shows that the mode of distribution of the u_0 velocity component accords with the requirements of Maxwell's Law.

Comparing (25) with (20) we see that $\dfrac{\nu\epsilon V_1}{4RTz_1^2} = 55\cdot1$. Substituting the known value of $\nu\epsilon$ and the experimental values of V_1, T, and z_1, this gives R $= 4\cdot8 \times 10^3$, which, considering all the possible sources of error, is in satisfactory agreement with the theoretical value $3\cdot71 \times 10^3$. Another test, which is not independent of the last, can be applied by plotting log i against x^2, when a straight line should be obtained in accordance with (21). This is found to be the case except for large values of x, when the currents are so small that various sources of error have a serious effect. The value of R obtained from the slope of the line thus got was found to be $4\cdot7 \times 10^3$. The third method is independent of the foregoing, and with the apparatus actually used should be less reliable. Since the total current from the slit and the plates is $N_1\epsilon$, if j_0 denotes the fraction of this which passes through the slit when in the symmetrical position, then

$$j_0 = i_0/N_1\epsilon, \qquad . \qquad . \qquad . \qquad (26)$$

and from (19)

$$R = \frac{1}{T}\frac{\nu\epsilon V_1}{4\pi z_1^2}\frac{\xi^2}{j_0^2}, \qquad . \qquad . \qquad (27)$$

where ξ is the width of the slit and j_0 is the maximum value of the ordinate in the left-hand curve of Fig. 18. On substituting the experimental values (27) gave R $= 2\cdot7 \times 10^3$. In applying the graphical method it was found that the points were not quite symmetrical on the two sides of the central position ($x = 0$). On one side they were very close to the curve $y = 1\cdot141x^2e^{-0\cdot0495x^2}$, and on the other to the curve $y = 1\cdot075x^2e^{-0\cdot0495x^2}$. These curves are of the form demanded by Maxwell's Law. If n is Avogadro's number the u_0 part of the kinetic energy was found, for this number of electrons, to be $3\cdot2 \times 10^6$ ergs. per c.c., as against the calculated value $2\cdot8 \times 10^6$ ergs. per c.c. The value of R calculated from the exponent $0\cdot0495x^2$, assuming Maxwell's Law to hold, was $5\cdot4 \times 10^3$.

These methods are not as accurate as the one used in testing the normal component of velocity; so that the mean of the four values obtained from R, namely $4 \cdot 4 \times 10^8$ instead of $3 \cdot 711 \times 10^8$, is to be regarded as satisfactory under the circumstances.

The other case in which (16) simplifies, arises when $V_1 = 0$, when it reduces to

$$i = 2N_1\epsilon \left(\frac{h^3 m^3}{\pi}\right)^{1/2} \frac{\xi}{z_1} \int_0^\infty w_0^2 e^{-hm\left(1 + \frac{x^2}{z_1^2}\right)w_0^2} dw_0$$

$$= N_1\epsilon \frac{\xi}{2} \frac{z_1^2}{(z_1^2 + x^2)^{3/2}} \qquad . \qquad . \qquad . \quad (28)$$

If, as before, we denote the current through the slit when $x = 0$ by i_0, then from (28)

$$i_0 = N_1\epsilon\xi/2z_1 ; . \qquad . \qquad . \qquad . \quad (29)$$

so that

$$i/i_0 = \left(1 + \frac{x^2}{z_1^2}\right)^{-3/2} \qquad . \qquad . \qquad . \quad (30)$$

Thus the ratio, of the current which flows through the slit at different distances x from the central position, to its value when $x = 0$, is determined solely by the distance z between the plates and is independent of the temperature of the source and the charge of the electrons. The extent to which this formula is confirmed by the observations is shown in Fig. 19, where the full line represents the curve calculated simply from the distance between the plates, and the points shown represent observations under different conditions as to the temperature of the platinum, the magnitude of the emission, and the direction of the heating current. A similar agreement was obtained when the distance between the plates was altered.

Taken in conjunction with the experiments on the distribution of the normal component of velocity described on p. 159, these experiments with zero electric field afford a valuable confirmation of the conclusion that Maxwell's Law of distribution holds good for the tangential components of velocity. For it is easily shown [1] that if Maxwell's Law holds for the normal component and not for the tangential, or vice

[1] O. W. Richardson, " Phil. Mag.," Vol. XVI, p. 909 (1908).

versa, the results of the experiments with zero electric field would be different from those obtained.

It will be noticed from the figure that the observed currents are consistently larger than the theoretical values at considerable distances from the central position. A similar deviation from theory, but usually more marked, is observed in the experiments in which an accelerating potential is applied between the plates. This difference which, although rather erratic in its behaviour, usually increases with continued heating of the strips, seemed to point to a deviation from Maxwell's Law of

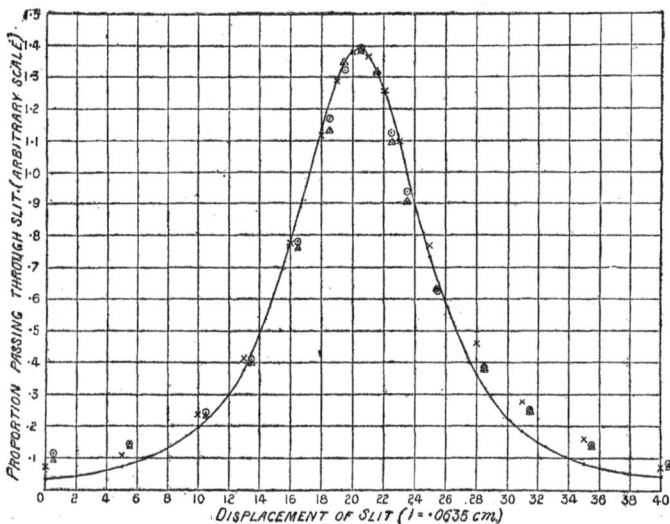

FIG. 19.

velocity distribution. A large number of experiments on the subject, however, led the writer to conclude that there was no foundation for such a view; but that the effects in question were due to subsidiary causes, such as the roughness of the metal surface caused by recrystallization, and the deflexion of the moving electrons by gas molecules. It may be, however, that these influences have been overestimated.

In all the experiments with a movable slit it was noticed that the current received by the slit was always greater than that received by an equal area of the plates when in the same position. This effect was attributed by the writer[1] to the reflexion of the electrons impinging on the plates. It was estimated that about 30 per cent of the slow-moving electrons present in the absence of an electric field were reflected in this way from a brass surface. About the same time, similar effects were observed by von Baeyer[2] in experiments with the electrons emitted from Wehnelt cathodes, and were ascribed by him to the same cause.

The kinetic energy of the electrons emitted by carbon and tungsten has recently been investigated by Schottky[3] who measured the electron current i which flowed from hot wires of circular section made of these materials, to a concentric cylindrical electrode, against a difference of potential V_1. If the initial distribution of velocity among the emitted electrons is in accordance with Maxwell's Law, and if r and R are the radii of the wire and cylinder respectively, then

$$i = i_0 \frac{2}{\sqrt{\pi}} \left\{ e^{-2he\,V_1} \int_0^{\sqrt{2he\lambda V_1}} e^{\,} \quad dx + \int_{\sqrt{2he\lambda V_1}}^{\infty} e^{-x^2} dx \right\} \quad (31)$$

where i_0 is the value of i when $V_1 = 0$, $\theta = r/R_1$, and $\lambda = (1 - \theta^2)^{-1}$. Under the conditions which held during the experiments $\left(\theta = r/R < 1/30 \text{ and } n = 2heV_1 = \dfrac{ve \cdot V_1}{RT} < 10 \right)$ equation (31) is identical, within $\frac{1}{2}$ per cent, with the equation

$$i = i_0 \frac{2}{\sqrt{\pi}} \left\{ e^{-n} \sqrt{n} + \int_{\sqrt{n}}^{\infty} e^{-x^2} dx \right\} . \qquad (32)$$

The experimental results were compared with the values calculated from (32), a value of n being assumed so as to give as close a fit as possible. In every case a fair agreement with the formula was obtained, provided the currents which reached the electrodes were small; but with larger currents, obtained either with higher temperatures of the hot wire, or with small applied potential differences, there were consistent deviations from the formula. This deviation is accounted for

[1] " Phil. Mag.," Vol. XVI, p. 898 (1908) ; Vol. XVIII, p. 694 (1909) ; " Phys. Rev.," Vol. X, p. 168 (1909).

[2] " Verh. d. Deutsch. Physik. Ges.," 10 Jahrg., pp. 96, 953 (1908) ; " Phys. Zeits.," 10 Jahrg., p. 168 (1909).

[3] " Ann. der Physik," Vol. XLIV, p. 1011 (1914).

satisfactorily by the effects arising from the mutual repulsion of the electrons discussed in Chapter III. From the experimental values of T and V_1 knowing $\nu\epsilon$, the value of R can be calculated from these experiments. The values found for the kinetic energies are all somewhat above the theoretical value, the error for carbon ranging from 5 per cent to 26 per cent, and for tungsten from 2 per cent to 23 per cent, but it is probably difficult to obtain the temperatures accurately under the conditions of the measurements.

An important change in the method of experimenting was introduced by v. Baeyer and adopted by Schottky, which removes two possible sources of error present in the earlier experiments. Both in the heating circuit and in the line for measuring the thermionic current, he inserted a make and break switch. These switches were both operated 250 times per second by the same mechanism, so that when one was in the other was. out. Thus, when the electron currents were being measured, there was no magnetic field and no fall of potential down the wire due to the heating current, and any error which might arise from their presence was, therefore, avoided. Owing to the short time of interruption of the current, the fall of temperature of the wire thereby arising would be inconsiderable.

It follows from the various experiments which have been described that the velocities of the electrons emitted by hot metals are identical with those which would be possessed by the molecules of a gas, of equal molecular weight with the electrons, which cross any area drawn in an enclosure containing the gas in equilibrium at the same temperature as the hot metal. Since the proof of this identity rests on experiment it can only be held to be established within the limitation of accuracy set by the experimental methods employed ; but the deviations from the strict theoretical requirements have always been found to be such as could readily be accounted for as arising from various secondary causes which it has not been possible completely to eliminate. From this result the application of Maxwell's Law of velocity distribution to the atmospheres of electrons in equilibrium outside metals follows immediately, but, as has been pointed out on p. 155, it does not necessarily apply to the electrons inside the metals.

It is obviously impossible to make experiments, similar to those described, with gases whose molecules are uncharged, on account of the smallness of the controllable forces which it is possible to bring to bear on individual molecules. For this reason the experiments of the writer and F. C. Brown formed the first experimental investigation of the distribution of velocity among the particles of any system to which Maxwell's Law could apply, although the law itself was predicted by Maxwell[1] on theoretical grounds in 1860.

Experiments carried out in the last few years under the direction of the writer by S. L. Ting indicate that the deviations from the requirements of Maxwell's Law in the direction of higher emission velocities recorded in the foregoing experiments may be real and not due to experimental inaccuracies. It appears that a very common distribution is one which satisfies Maxwell's Law, except that the average energy or temperature of the electrons greatly exceeds that of the metal from which they originate. In the majority of experiments made by Ting with tungsten and platinum the measured average energy approached twice the value expected from the temperature of the metal on the classical kinetic theory. There seems no reason for doubting the reliability of these results, but owing to the far-reaching consequences which their validity would entail, it is desirable that they should be subjected to still further critical examination. It may, however, be remarked that the effects observed so far are not unlike what might be expected on the quantum type of theory developed in Chapter II, p. 37. Until the broad facts to be explained are explored in more detail I have thought it better not to make any substantial alterations in the preceding and following accounts of these phenomena which were built up on the hypothesis that Maxwell's Law held accurately. In so far as the distribution turns out to be of the same form as Maxwell's, but with a different value for the average energy constant, the formulæ will still hold if we interpret the tempera-

[1] " Phil. Mag.," Vol. XIX, p. 22 (1860).

ture T, not as that of the metal but as some different temperature which is characteristic of the emitted electrons. It appears necessary to assume that the energy of the emitted electrons as measured does not represent that appropriate to a true equilibrium configuration for a dilute electron atmosphere; for, in equilibrium, the temperature of such an atmosphere could not differ from that of the emitting metal without contravening the second law of thermodynamics. It is also possible that these deviations from Maxwell's law are caused by gas effects or some action of surface layers which is not yet understood.

2. Steady Thermionic Currents between Conductors Maintained at Definite Temperatures and Potentials.

The case considered on p. 158 of the electron current from a hot strip to a neighbouring slit forms an example of a class of problems which the writer[1] has shown can be solved in a much more general manner. Suppose that in a region of space otherwise vacuous there is a hot surface A emitting ions and one or more conducting surfaces B. There may be an electric field in the region under consideration; so that any or all of the surfaces may be charged. The ions emitted by the surface A will move under the combined influence of their initial velocity and of the electric field and will ultimately either return to A, reach B, or go off to an infinite distance. If the distribution of temperature on the surface A is maintained constant the number, and mode of distribution of velocity, of the ions it emits will remain constant, and if in addition the potentials of the various surfaces are maintained constant, it is clear that, whatever may happen at first, a steady state will ultimately be established in which the number and mode of distribution of velocity among the ions received by any of the surfaces in a given time will be invariable. The problem is to find the number of ions which reach any of the surfaces B in a given time, together with their velocity components, when

---

[1] " Phil. Mag.," Vol. XVII, p. 813 (1909).

the steady state has been established. In the discussion it will be assumed that the motion of the ions is determined *solely* by their positional and velocity co-ordinates at emission and by the electric field. The forces exerted by the ions on each other[1] and by molecules of gas into whose spheres of action they may chance to penetrate are left out of account. These conditions are realized if thermionic currents of moderate size are experimented with in high vacua. In order to avoid complications arising out of recombination we shall also suppose the temperature conditions to be such that ions of one sign only occur. We shall now consider the general problem, using rectangular co-ordinates.

Let the co-ordinates of a point of the surface A be $x_0 y_0 z_0$ and let an ion be projected from $x_0 y_0 z_0$ with the velocity components $u_0 v_0 w_0$. Let us seek the condition that this shall strike the surface B, whose equation is

$$\psi(xyz) = 0, \qquad . \qquad . \qquad . \qquad (33)$$

within an infinitesimal distance of the point $x_1 y_1 z_1$. If V is the potential at any point of the field, the equations of motion of the ion will be

$$m\frac{\partial^2 x}{\partial t^2} = -\epsilon\frac{\partial V}{\partial x}, \; m\frac{\partial^2 y}{\partial t^2} = -\epsilon\frac{\partial V}{\partial y}, \; m\frac{\partial^2 z}{\partial t^2} = -\epsilon\frac{\partial V}{\partial z} \qquad (34)$$

On integration these equations give three equations between x, y, z, and t involving six arbitrary constants which are determined by the values of $x_0 y_0 z_0 u_0 v_0 w_0$. After elimination of the time there result two equations which may be written

$$\phi_1(xyzx_0 y_0 z_0 u_0 v_0 w_0) = 0 \qquad . \qquad . \qquad . \qquad (35)$$
$$\phi_2(xyzx_0 y_0 z_0 u_0 v_0 w_0) = 0 \qquad . \qquad . \qquad . \qquad (36)$$

The curve in which the surfaces ϕ_1 and ϕ_2 intersect is the trajectory of the particle projected under the given initial conditions. The intersection of this curve with the surface $\psi(xyz) = 0$ gives the point where the particle strikes the surface. The co-ordinates $x_1 y_1 z_1$ of such points will therefore be given by solving (33), (35), and (36) for x, y, and z, and

---

[1] Particular problems of the same general character in the treatment of which the influence of the interionic forces have been taken into account have been considered on p. 47 and p. 70.

the density of these points on the surface ψ will determine the thermionic current density into this surface in the steady state.

In general the equations for $x_1 y_1 z_1$ will not be of the first degree; so that there will be a number of roots corresponding to the successive real and imaginary intersections of the surfaces ϕ_1, ϕ_2, and ψ. In any case the path of the particle will end as soon as it has reached the conducting surface B, and if this surface includes the whole of the analytical surface $\psi (x_1 y_1 z) = 0$ the root to be chosen is that real root which corresponds to the shortest time of transit from $x_0 y_0 z_0$. The proper root can usually be picked out in simple cases. If the surface B is only a part of the analytical surface $\psi = 0$ bounded by a curve or curves, it may in general be necessary to include roots corresponding to any number, less than that of the degree of the equations, of previous intersections of the trajectory and the surface $\psi = 0$. The problem then becomes much more complicated.

The equations (35) and (36) may be solved for u_0 and v_0 giving

$$u_0 = \phi_3(xyz x_0 y_0 z_0 w_0) \qquad . \qquad . \qquad . \quad (37)$$
$$v_0 = \phi_4(xyz x_0 y_0 z_0 w_0) \qquad . \qquad . \qquad . \quad (38)$$

The equation $\phi_3 = $ constant, together with $\psi(xyz) = 0$, will determine a curve lying in the surface ψ which contains the points of intersection with ψ of all trajectories for which u_0 and w_0 are constant. Similarly $\phi_4 = $ constant determines a curve corresponding to constant values of v_0 and w_0. If ξ and η denote lengths laid out along the normals to the level surfaces of u_0 and v_0, respectively, at any point, then

$$\left.\begin{array}{l} \dfrac{\partial u_0}{\partial \xi} = \sqrt{\left(\dfrac{\partial \phi_3}{\partial x}\right)^2 + \left(\dfrac{\partial \phi_3}{\partial y}\right)^2 + \left(\dfrac{\partial \phi_3}{\partial z}\right)^2} \\[3mm] \dfrac{\partial v_0}{\partial \eta} = \sqrt{\left(\dfrac{\partial \phi_4}{\partial x}\right)^2 + \left(\dfrac{\partial \phi_4}{\partial y}\right)^2 + \left(\dfrac{\partial \phi_4}{\partial z}\right)^2} \end{array}\right\} \quad . \quad (39)$$

Let the number of particles which are emitted in unit time with velocity components between u_0 and $u_0 + du_0$ be denoted by $f_1(u_0)du_0$, the corresponding number with respect to v_0 and $v_0 + dv_0$ being $f_2(v_0)dv_0$. For a constant value of w_0 the number which simultaneously have velocity components within the ranges above will be proportional to $f_1(u_0)f_2(v_0)du_0 dv_0$ and these will fall on an area dS of the surface $\psi = 0$, given by

$$dS \cos(u_0 v_0 nS) = \frac{du_0\, dv_0}{\dfrac{\partial u_0}{\partial \xi}\dfrac{\partial v_0}{\partial \eta}} \times \frac{1}{\sin u_0 v_0},$$

where $u_0 v_0 nS$ is the angle between the normal to the surface $\psi = 0$ and the tangent to the curve in which the surfaces $u_0 = \phi_3$, $v_0 = \phi_4$ intersect; and $u_0 v_0$ is the angle between the normals to the surfaces $u_0 = \phi_3$ and $v_0 = \phi_4$. Hence,

$$du_0 dv_0 = \frac{\left(\dfrac{\partial \phi_3}{\partial y_1}\dfrac{\partial \phi_4}{\partial z_1} - \dfrac{\partial \phi_3}{\partial z_1}\dfrac{\partial \phi_4}{\partial y_1}\right)\dfrac{\partial \psi}{\partial x_1} + \left(\dfrac{\partial \phi_3}{\partial z_1}\dfrac{\partial \phi_4}{\partial x_1} - \dfrac{\partial \phi_3}{\partial x_1}\dfrac{\partial \phi_4}{\partial z_1}\right)\dfrac{\partial \psi}{\partial y_1} + \left(\dfrac{\partial \phi_3}{\partial x_1}\dfrac{\partial \phi_4}{\partial y_1} - \dfrac{\partial \phi_3}{\partial y_1}\dfrac{\partial \phi_4}{\partial x_1}\right)\dfrac{\partial \psi}{\partial z_1}}{\left[\left(\dfrac{\partial \psi}{\partial x_1}\right)^2 + \left(\dfrac{\partial \psi}{\partial y_1}\right)^2 + \left(\dfrac{\partial \psi}{\partial z_1}\right)^2\right]^{1/2}}\, dS$$

(40)

$$= \chi(x_1 y_1 z_1 x_0 y_0 z_0 w_0)dS \quad . \quad . \quad . \quad (41)$$

If the probability that the w_0 component of velocity lies between w_0 and $w_0 + dw_0$ is denoted by $f_3(w_0)dw_0$, the number of electrons reaching the surface $\psi = 0$ with values of w_0 within this range is proportional to

$$dw_0 \int \int f_3(w_0)f_1(u_0)f_2(v_0)\chi dS,$$

and, if N is the total number of ions emitted in unit time by unit area of the surface A, the total number N_B received by the surface ψ will be the real part of

$$\int \int N dS_0 \int dw_0 \int \int f_3(w_0)f_1(\phi_3)f_2(\phi_4)\chi dS, \quad . \quad (42)$$

where dS_0 denotes an element of the surface A and the integral with respect to dw_0 is taken over all the values of w_0 which occur.

If we multiply (42) by the charge ϵ of an ion we obtain the current to the surface ψ. We can obtain the three components of the resultant pressure on this surface due to the impact of the ions if we multiply the integrand with respect to dS by $m\dfrac{\partial x_1}{\partial t}$, $m\dfrac{\partial y_1}{\partial t}$, and $m\dfrac{\partial z_1}{\partial t}$ respectively. The values of the velocities are obtained from equations (34) and should be expressed as functions of $x_1 y_1 z_1 x_0 y_0 z_0$ and w_0 by means of the equation previously given. In a similar way we obtain the kinetic energy received by the surface if we multiply the integrand by

$$\tfrac{1}{2}m\left[\left(\frac{\partial x_1}{\partial t}\right)^2 + \left(\frac{\partial y_1}{\partial t}\right)^2 + \left(\frac{\partial z_1}{\partial t}\right)^2\right].$$

This must be identical with (42) $\times$ $\epsilon(V_0 - V_1)$ + the value of the integral when $\tfrac{1}{2}m(u_0^2 + v_0^2 + w_0^2)$ is substituted for

$$\tfrac{1}{2}m\left[\left(\frac{\partial x_1}{\partial t}\right)^2 + \left(\frac{\partial y_1}{\partial t}\right)^2 + \left(\frac{\partial z_1}{\partial t}\right)^2\right],$$

V_0 being the potential of the surface A and V_1 that of ψ.

It is often easier to effect a direct integration with respect to u_0 and v_0 than to carry out the transformation outlined above. Since $\chi dS = du_0 dv_0$ (42) may be replaced by

$$\int\int N dS_0 \int dw_0 \int\int f_3(w_0) f_1(u_0) f_2(v_0) du_0 dv_0, \qquad . \quad (43)$$

the limits of integration being suitably changed. If the surface B forms the whole of the analytical surface $\psi(xyz) = 0$ the limits of integration for u_0 and v_0 will be determined, for any value of w_0, by the values of u_0 and v_0 which correspond to the curve which is the locus of the points at which the trajectories having the given value of w_0 are tangential to the surface $\psi(xyz) = 0$. They will thus be certain functions of w_0 which are determined by the equation to the surface. If the surface B consists of the portion of $\psi = 0$ which is cut off by some closed curve, the limits for u_0 and v_0 will be determined partly by the bounding curve and partly by the locus of the tangents. It will often be possible so to choose the direction of w_0 that $f_3(w_0)$ does not depend on u_0 and v_0.

THE INITIAL VELOCITIES.

The experiments described at the beginning of this chapter showed that the initial velocities of the electrons were distributed in accordance with Maxwell's Law. We shall see later that the same statement has been found to be true for the positive ions emitted from hot bodies in a large number of cases (cf. p. 206). We can therefore write down the functions $f_1(u_0)$, $f_2(v_0)$, and $f_3(w_0)$ which express the initial frequency of a velocity component within a given range. They will depend both on the kind of axes chosen and on their orientation relative to the emitting surface. The following list embraces all the more important cases. In each case N is the total number of ions emitted per unit area in the interval of time under consideration, m is the mass of an ion and $3/4h$ the mean kinetic energy.

1. *Rectangular Co-ordinates.*—The formulæ for this case are repeated here for the sake of completeness although they have already been given. The axis of $\dot{z}$ is normal to the emitting surface.

Number between $\dot{z}$ and $\dot{z} + d\dot{z} = \mathrm{N}\dot{z}\mathrm{F}(\dot{z})d\dot{z} = 2\mathrm{N}hm\dot{z}\ e^{-hm\dot{z}^2}d\dot{z}$ (44)

$$\left.\begin{array}{l} \text{,,}\qquad\text{,,}\qquad \dot{x}\ \text{,,}\ \dot{x} + d\dot{x} = \mathrm{N}f(\dot{x})d\dot{x} = \mathrm{N}\left(\dfrac{hm}{\pi}\right)^{1/2}e^{-hm\dot{x}^2}d\dot{x} \\[2mm] \text{,,}\qquad\text{,,}\qquad \dot{y}\ \text{,,}\ \dot{y} + d\dot{y} = \mathrm{N}f(\dot{y})d\dot{y} = \mathrm{N}\left(\dfrac{hm}{\pi}\right)^{1/2}e^{-hm\dot{y}^2}d\dot{y} \end{array}\right\} \quad (45)$$

2. *Spherical Co-ordinates.*—Let $\dot{\psi}$ be the resultant velocity, θ the angle it makes with the normal to the surface, and ϕ the angle the plane containing $\dot{\psi}$ and the normal makes with a fixed plane containing the normal. Then the number emitted per unit area per second which have $\dot{\psi}$ between $\dot{\psi}$ and $\dot{\psi} + d\dot{\psi}$, θ between θ and $\theta + d\theta$, and ϕ between ϕ and $\phi + d\phi$ is

$\mathrm{N}\dot{\psi}\cos\theta\mathrm{F}(\dot{\psi}\cos\theta)f(\dot{\psi}\sin\theta\cos\phi)f(\dot{\psi}\sin\theta\sin\phi)\ \dot{\psi}^2d\dot{\psi}\sin\theta d\theta d\phi$

$= \mathrm{N}\dot{\psi}^3\mathrm{F}(\dot{\psi}\cos\theta)\,\mathrm{F}_1(\dot{\psi}\sin\theta)\sin\theta\cos\theta d\dot{\psi} d\theta d\phi$

$= \dfrac{2h^2m^2}{\pi}\dot{\psi}^3e^{-hm\dot{\psi}^2}\sin\theta\cos\theta d\dot{\psi} d\theta d\phi$.   .   .   .   (46)

3. *Cylindrical Co-ordinates.*—(a) The axis of z is along the normal to the surface, ρ is the radius perpendicular to the axis of z, and θ is the angle $\dot{\rho}$ makes with a fixed plane passing through the z axis.

The number between $\dot{z}$ and $\dot{z} + d\dot{z} = \mathrm{N}\dot{z}\mathrm{F}(\dot{z})d\dot{z} =$

$$2\mathrm{N}hm\dot{z}e^{-hm\dot{z}^2}\ d\dot{z} \qquad . \qquad . \qquad . \quad (47)$$

whilst the number for which $\dot{\rho}$ is between $\dot{\rho}$ and $\dot{\rho} + d\dot{\rho}$ and θ simultaneously between θ and $\theta + d\theta$ is

$$\mathrm{N}f(\dot{\rho}\sin\theta)f(\dot{\rho}\cos\theta)\,d\dot{\rho}\,\dot{\rho}d\theta = \mathrm{N}\dfrac{hm}{\pi}\,\dot{\rho}e^{-hm\dot{\rho}^2}d\dot{\rho}d\theta \quad . \quad (48)$$

(β) The axis of z lies in the tangent plane to the surface.

$\dot{\phi}$ is the total component of velocity perpendicular to z, i.e. the projection of the resultant velocity on a plane perpendicular to the z axis. θ is the angle $\dot{\phi}$ makes with the plane containing the axis of z and the normal to the surface. The number whose velocity components lie between z and $z + dz$ is

$$N f(\dot{z}) d\dot{z} = N \left(\frac{hm}{\pi}\right)^{1/2} e^{-hm\dot{z}^2} d\dot{z} \qquad . \qquad . \quad (49)$$

The number which have components between $\dot{\phi}$ and $\dot{\phi} + d\dot{\phi}$, and for which at the same time θ lies between θ and $\theta + d\theta$, is

$$N\dot{\phi} \cos \theta F(\dot{\phi} \cos \theta) f(\dot{\phi} \sin \theta) \dot{\phi} d\dot{\phi} d\theta$$

$$= 2N\left(\frac{h^3 m^3}{\pi}\right)^{1/2} \dot{\phi}^2 e^{-hm\dot{\phi}^2} \cos \theta d\dot{\phi} d\theta \quad . \qquad . \quad (50)$$

The number for which $\dot{\phi}$ lies between $\dot{\phi}$ and $\dot{\phi} + d\dot{\phi}$ and for which θ has any value will therefore be

$$N\dot{\phi}^2 d\dot{\phi} \int_{-\pi/2}^{\pi/2} F(\dot{\phi} \cos \theta) f(\dot{\phi} \sin \theta) \cos \theta d\theta$$

$$= 4N\left(\frac{h^3 m^3}{\pi}\right)^{1/2} \dot{\phi}^2 e^{-hm\dot{\phi}^2} d\dot{\phi} \qquad . \qquad . \quad (51)$$

In the paper by the writer referred to on p. 173 from which the matter in this section is practically an excerpt, the general solution is applied to a number of particular cases. Although the results are of considerable importance, it would take up too much space to do much more than enumerate the problems considered. The reader who is especially interested in this part of the subject may be referred for details to the original paper, where the following particular cases are considered : (1) No electric field between A and B. (2) A and B are portions of parallel planes, and the electric intensity is uniform and normal to the planes. When A and B are narrow parallel strips of indefinite length this case becomes the same as that considered on p. 158, and the solution by the general method is found to be identical with that given on p. 159. The equations (37) and (38) for u_0 and v_0 respectively are quadratic. By taking the positive sign we obtain the first intersection of the trajectory with the plane B and by taking the negative sign the second intersection. Taking the second intersection, and making the plane B coincide with the plane A, we obtain an expression for the current emitted by one part of a plane and returned to it at another part in a retarding field. (3) A and B are inclined planes, and the electric intensity is uniform and normal to A. In this case again we obtain the number of ions which return to A in a retarding field, by taking the second intersection and rotating the plane of B until it becomes coincident with that of A. (4) A is a circular cylinder surrounded by a thick-walled tube C in which a narrow gap is cut perpendicular to the axis of the tube. The problem is to find the number of ions which pass through the gap and reach an outer concentric cylinder B, when A and C are at the same potential which is different from that of B.

The case in which A and B are coaxial circular cylinders of indefinite length maintained at a constant difference of potential which retards the ions passing from A to B has been considered by Schottky.[1] The solution is given on p. 170 where we saw also that it had been confirmed by experiment.

3. The Latent Thermal Effects.

In this section in order to attain clearness in the interpretation of the various factors which determine the magnitude of the effects it is necessary to adopt some hypothesis as to the magnitudes of the kinetic energy of the electrons inside and outside metals. The following account is based on the classical kinetic theory of metallic conduction. If a theory of the quantum type should prove to be in better agreement with the facts various modifications in the formulæ or in their interpretation become necessary. Most of these are taken care of if we interpret the quantities T and T_0 as the temperatures appropriate to the kinetic energies of the external and internal electrons respectively on the equipartition hypothesis. Fortunately the terms in T and T_0 are small in any event and the precise signification attributed to them does not seriously affect the broad interpretation of the results.

[1] "Ann. der Physik," Vol. XLIV, p. 1011 (1914).

Loss of Energy due to Electron Emission.—We have just seen that when electrons escape from a hot body they carry with them on the average the definite amount of kinetic energy $2k\mathrm{T}$, where T is the temperature of the hot body. In addition, we saw in Chapters II and III that, in order to escape, each electron had to do an amount of work w against the forces tending to retain it in the interior of the substance. It follows that, for each electron which escapes, the hot body will suffer a loss of energy equal to $\phi + 2k\mathrm{T}$, and, if i is the thermionic current to the hot body, this surface loss of energy [1] will amount, per unit time, to

$$U = \frac{i}{e}(w + 2k\mathrm{T}) \qquad . \qquad . \qquad . \quad (52)$$

$$= i\left(\phi + \frac{2\mathrm{R}}{ve}\mathrm{T}\right) \qquad . \qquad . \qquad . \quad (53)$$

where ϕ is the potential difference through which an electron has to fall in order to acquire an amount of energy equal to w.

The loss of energy under consideration is analogous to the heat lost during the evaporation of liquids, and it may, in fact, be regarded as the latent heat of evaporation of electricity from the substance in question. On account of the very rapid increase of i with rising temperature the heat lost in this way will also increase with corresponding rapidity. With substances like carbon and tungsten this loss of energy should become equal to, and ultimately exceed, that arising from electro-magnetic thermal radiation at temperatures below 3000° C. It is to be borne in mind that energy will only be lost in this way so long as the electrons are emitted. In the case of an insulated hot body it will soon cease, as the emission of electrons is stopped by the positive charge it leaves on the hot body.

The first experiments to detect and measure this effect were made by Wehnelt and Jentzsch [2] using the emission from lime-coated platinum wires. The wire formed one of the two low-resistance arms of a Wheatstone's bridge circuit through

---

[1] O. W. Richardson, " Phil. Trans., A.," Vol. CCI, p. 497 (1903).

[2] "Verh. der Deutsch. Physik. Ges.," 10 Jahrg., p. 610 (1908); " Ann. der Physik," Vol. XXVIII, p. 537 (1909).

which a large heating current flowed in the usual way (p. 17). The resistance and therefore the temperature of the wire could be kept very accurately constant by controlling the external regulating resistances. The main current also flowed through a suitable standard resistance. The potential drop along this was measured by a sensitive potentiometer arrangement which enabled extremely small variations of the heating current to be determined. It was found that when the hot wire was charged negatively, so as to cause the electron current to flow from it, it was necessary to increase the magnitude of the heating current in order to maintain the resistance of the wire constant. If R_1 is the resistance of the wire, i_1 the value of the heating current when the thermionic current is not flowing, and $i_1 + di_1$ the value required to maintain the resistance at R_1 when the thermionic current is flowing, the rate of supply of additional energy necessary to keep the temperature of the wire constant is

$$U = R_1[(i_1 + di_1)^2 - i_1^2] = 2R_1 i_1 di_1 \quad . \quad . \quad (54)$$

neglecting $R_1 di_1^2$. If this energy is entirely used up in counteracting the cooling due to the emission of the electrons, and if there are no subsidiary disturbing effects, we see from equation (53) that U will also be equal to $i [\phi + \dfrac{2R}{\nu e} (T - T_0)]$,

where i is the thermionic current and T_0 the temperature of the cold part of the system. The term in T_0 is added because the electrons carry the corresponding quantity of energy when they flow into the wire at the cold ends. In the equation

$$i[\phi + \frac{2R}{\nu e} (T - T_0)] = 2R_1 i_1 di_1 \quad . \quad . \quad (55)$$

all the quantities are known or measurable except ϕ ; so that these experiments should enable ϕ to be determined. Unless special precautions are taken there are in experiments of this character a number of possible disturbing phenomena which may seriously affect the results. The most important of these arise from the direct action of the thermionic current in upsetting the balance of the Wheatstone's bridge, and in modifying the distribution of temperature along the hot wire. The conditions which have to be satisfied in order either to eliminate the effects of these disturbing actions or to make them so small as to be innocuous are discussed in a paper by H. L. Cooke and the writer.[1]

Working with the method outlined, Wehnelt and Jentzsch were able to show that the emission of electrons caused a cooling of the wire at low temperatures. This changed to a heating effect at high temperatures and with large thermionic currents. The heating effect was satisfactorily attributed to the energy communicated to the wire by the positive ions liberated, by impact ionization, from the gas evolved by the wires when strongly heated. On the other hand, the phenomena at low temperatures were not in accordance with the requirements of the theory. At the lowest temperature (950° C.) the value of ϕ found was about ten times as great as that deduced from the temperature variation of the electron emission from lime, and instead of being constant it diminished rapidly with rising temperature. Similar results have since been obtained by Schneider.[2] It now appears that lime was chosen unfortunately for the purpose of investigating this effect, as its behaviour presents abnormalities which are not exhibited by the highly refractory metals.

The first clear proof of the existence of the cooling effect predicted by the theory was given by H. L. Cooke and the writer [3] as a result of experiments made with osmium filaments. In these experiments Wehnelt and Jentzsch's method was modified somewhat; the change in the resistance R_1, due to turning the thermionic current i off and on, was observed when the heating current i was kept constant. This simplifies the manipulation very considerably, although the numerical reduction of the results becomes rather more complicated. Precautions were taken also to ensure the absence of errors arising from the effect of the thermionic current itself on the

---

[1] "Phil. Mag.," Vol. XXV, p. 628 (1913); cf. also *ibid.*, Vol. XX, p. 173 (1910).

[2] "Ann. der Physik," Vol. XXXVII, p. 569 (1912).

[3] "Phil. Mag.," Vol. XXV, p. 624 (1913).

Wheatstone's bridge galvanometer and from the alteration in the distribution of temperature along the filament due to the Joule effect of the thermionic current. For these matters the original paper must be consulted. It may be permissible to point out that a factor $1/i_1$ has been omitted from the right-hand side of equation (13), p. 635, and that the same error has been copied into a later paper on a similar subject [equation (13'), "Phil. Mag.," Vol. XXVI, p. 475 (1913)]. In all, 37 determinations of ϕ were made under conditions as varied as possible. The experiments involved the following range of variation of the quantities entering into the reduction formula : the thermionic current i from 2×10^{-5} to 8×10^{-4} amp., the heating current i_1 from 0·430 to 0·687 amp., the resistance R_1 from 4·216 to 5·533 ohms, and the potential driving the thermionic current from 12·1 to 24·4 volts. All the thirty-seven resulting values of $\phi + 2\dfrac{R}{\nu e}(T - T_0)$ fell between 4·16 and 6·16 volts, and if five of them, which involved the measurement of extremely small deflexions and are therefore liable to large observational errors, are neglected, all the remaining thirty-two lie between 4·59 and 5·36 volts. Thus the results are quite consistent and the evidence is definite that there is no large variation of ϕ with T. The mean of all the thirty-seven values gives

$$\phi = 4·7 \text{ equivalent volts.}$$

The value of b corresponding to this[1] would be

$$b = 54·600 \text{ degrees.}$$

Since the chemical properties of osmium are similar to those of platinum, it is satisfactory to note that this value of b falls within the limits of the platinum values on p. 81, Chapter III.

Measurements of the same kind were also made by H. L. Cooke and the writer[2] using tungsten filaments mounted in tubes which had been very carefully treated to eliminate gaseous contamination. Six measurements gave values of ϕ showing an extreme variation of 0·8 volt, the mean value, after

---

[1] Deduced from the relation $\Phi = 8·59 \times 10^{-5} b$; cf., however, below.
[2] "Phil. Mag.," Vol. XXVI, p. 472 (1913).

correction for the error in equation (13′) of the paper already alluded to, being

$$\phi = 4\cdot63 \text{ equivalent volts.}$$

As the data available in this case are more definite than for osmium it is worth while to consider the precise meaning of ϕ a little more closely. As used in the present section ϕ denotes the excess, over the equilibrium value, of the kinetic energy, expressed in equivalent volts, which an electron inside the metal has to lose in order to escape with the velocity, and the kinetic energy, zero. In Chapter II (for example on p. 30) the symbol ϕ has been used with a different meaning. It there denotes the change in ergs in the energy of the system which takes place when an electron escapes from the hot body *under actual equilibrium conditions at temperature T.* Apart from the difference of dimensions which may be regarded as accidental, there is thus an important distinction between the two quantities. To distinguish between them we shall in this section denote the ϕ of Chapter II by Φ. Now Φ can be regarded as made up of three parts: (1) Φ_1 the change in potential energy at the interface, (2) $- K_1$ the internal kinetic energy, and (3) $+ K_2 = \frac{3}{2} kT$ the external kinetic energy.
Thus

$$\Phi = \Phi_1 - K_1 + \tfrac{3}{2} kT \quad . \quad\quad . \quad\quad . \quad (56)$$

In Chapter II, p. 32 *et seq.*, it is shown to follow from thermodynamics and the magnitude of the Thomson effect in metals that Φ is very nearly of the form

$$\Phi = \Phi_0 + \tfrac{3}{2} kT, \quad\quad . \quad\quad . \quad\quad . \quad (57)$$

where Φ_0 varies very little with T. It follows from (56) that $\Phi_1 - K_1$ varies very little with T. Again, the smallness of the specific heats of metals, as well as the quantum theory considerations in Chapter II, indicates that the energy K_1 of the internal electrons varies little if at all with T ; so that, to the same extent, the same must be true of Φ_1. Except for the difference of dimensions, which we may disregard for the moment, the ϕ of this section is evidently

$$\phi = \Phi_1 - K_1 + 2kT_0 = \Phi_0 + 2kT_0, \quad . \quad (58)$$

the term in T_0 appearing because we have already subtracted the small quantity $2kT_0$ in deducing ϕ. This inclusion is clearly undesirable except from the standpoint of the theory based on the classical dynamics; but in any event it is unimportant, as the amount so added is comparable with the errors of measurement. If we disregard it

$$\phi = \Phi_0 \quad . \quad . \quad . \quad . \quad (59)$$

From Chapter II, equation (17), $i = AT^2 e^{-\Phi_0/T}$; so that ϕ when corrected for the difference in dimensions is the numerator in the exponent in this equation. From equation (6) Chapter III, p. 64, we see that the corresponding value of the exponent b is

$$b = \Phi_0 + \tfrac{3}{2}T . \quad . \quad . \quad . \quad (60)$$

Corresponding to $\phi = 4 \cdot 63$ equivalent volts, $\Phi_0 = 5 \cdot 33 \times 10^4$ degrees and

$$b = 5 \cdot 62 \times 10^4 \text{ degrees} . \quad . \quad . \quad (61)$$

This is certainly a very satisfactory agreement with the values of b for tungsten deduced from the variation of the saturation current with temperature according to the formula $i = AT^{\frac{1}{2}}e^{-b/T}$. Langmuir's values for b under the best vacuum conditions vary from $5 \cdot 25 \times 10^4$ to $5 \cdot 58 \times 10^4$, the lower values being considered most satisfactory.

Cooke and Richardson[1] also made experiments on the cooling effect with the Wehnelt cathode, and were able to confirm the conclusion of Wehnelt and Jentzsch that the behaviour of these cathodes did not conform to the theory. Recently the action of this source of electrons has been examined by Wehnelt and Liebreich.[2] They find that, starting with a freshly prepared coating of lime and platinum, the emission diminishes with time on first heating, reaches a minimum, rises again, remains fairly constant for some time, increases sharply to a relatively high maximum and then falls away to small values. The precise character of these changes depends to a considerable extent on the temperature of the cathode and on the applied voltage, but they were found to be accompanied by corresponding changes in the magnitude of the

---

[1] " Phil. Mag.," Vol. XXVI, p. 472 (1913).

[2] " Verh. der Deutsch. Physik. Ges.," 15 Jahrg., p. 1057 (1913); " Physik. Zeits.," 15 Jahrg., p. 548 (1914).

cooling effect. When the emission was small so was the cooling effect, and vice versa. The measured values of

$$\phi + 2 \frac{R}{ve} (T - T_0)$$ varied between the extreme limits 2·24

and 10·66 equivalent volts. To explain these variations Wehnelt and Liebreich assume that in addition to the cooling due to the emission of electrons two other effects are present. These effects, which vary in magnitude with the duration of heating and with other conditions, are : (1) a heating effect due to the energy of positive ions received by the cathode, and formed by impact ionization in the gas liberated from it, and (2) a cooling effect arising from the volatilization of the lime which is partly enhanced by the positive ion bombardment. These assumptions are shown to give a satisfactory account, not only of the variations of the apparent experimental value of ϕ, but also of the concomitant variations of the saturation current with lapse of time under different applied potentials. The peculiar behaviour of lime cannot therefore be regarded as an argument against the general theoretical position.

The various difficulties which earlier investigators had experienced with Wehnelt cathodes have been overcome by W. Wilson [1] who used the standard oxide coated filaments prepared by the Western Electric Company. The tubes were very thoroughly freed from gases and the values of ϕ from the cooling effect compared with the corresponding quantity bk/e deduced from the temperature variation of the electron emission. The results for filaments covered with three different types of oxide coating are shown in the table on the following page. The data are very consistent and hold good for a given filament over long periods of time. The values of ϕ agree with those for bk/e within the limits of experimental error.

Wehnelt and Liebreich [2] also investigated the cooling effect from platinum alone and found values for ϕ varying between 5·78 and 6·04 equivalent volts. If we take 5·9 as the

[1] Cf. H. D. Arnold, " Phys. Rev.," Vol. XVI, p. 78 (1920).
[2] " Verh. der Deutsch. Physik. Ges.," loc. cit.

Composition of Oxide.	$\frac{bk}{e}$ (Volts).	ϕ (Volts).
BaO 50 per cent., SrO 50 per cent. . . .	2·02	1·97
	2·16	2·28
BaO 50 per cent., SrO 25 per cent., CaO 25 per cent.	2·34	2·39
	2·59	2·54
CaO	3·28	3·22
	3·49	3·51

mean value for ϕ, and calculate the corresponding value for the constant b in the emission formula in the same way as was done with tungsten on p. 185, we find

$$b = 7 \cdot 1 \times 10^4 \text{ degrees.}$$

This number is not far from the best values of b for platinum in the table on p. 81. Thus we see that for the metals tungsten and platinum, and probably also osmium, as well as from the Wehnelt cathode, the values of b calculated from the cooling effect are in agreement with those calculated from the temperature variation of the saturation currents.

H. H. Lester[1] has published a consistent series of determinations of ϕ which he has made with molybdenum, carbon, tantalum, and tungsten, from measurements of the cooling effect, using the same method as Cooke and Richardson. The particulars are collected in the following table :—

Substance.	Number of Measurements.	Extreme Values of ϕ (in Equivalent Volts).	Mean Value of ϕ (in Equivalent Volts).
Molybdenum	4	4·464—4·679	4·588
Carbon	14	4·140—4·97	4·55
Tantalum	7	4·308—4·734	4·511
Tungsten	10	4·190—4·724	4·478

The value for tungsten is slightly lower than that found by Cooke and Richardson (p. 185). The most interesting

[1] " Phil. Mag.," Vol. XXXI, p. 197 (1916).

feature of these results is that they make the values of ϕ practically identical for all the elements tested. If it could be established generally that ϕ is the same for all substances most important consequences would follow, amongst others the absence of contact electromotive force under good vacuum conditions.

4. THE HEAT LIBERATED DURING ELECTRON ABSORPTION.

When a powerful stream of slowly moving electrons is absorbed by a metal there is a liberation of heat which is the converse effect to that just considered. Naturally, if the electrons have been allowed to fall through any considerable difference of potential they will have acquired a corresponding amount of kinetic energy and this will appear in the form of heat on absorption. But the effect now under consideration occurs even when the stream of electrons reaches points just outside the absorbing surface with zero kinetic energy. It is caused by the work done on the electrons as they cross the surface layer, and may be regarded as analogous to the latent heat liberated during the condensation of vapours.

This effect has been investigated by the writer and H. L. Cooke.[1] Two short osmium filaments heated to a high temperature were used to supply the copious streams of electrons necessary. The metals to be tested, in the form of very thin strips, were wound on a very light glass frame so as to expose as much surface as possible. The frame was insulated from the osmium filaments and suitably mounted between them. By applying various small differences of potential the electrons could be directed from the osmium on to the metal strip. The strip formed one arm of a very sensitive Wheatstone's bridge, and the effect was detected and measured by the change of resistance and temperature experienced by the strip as the electrons were being absorbed. The strip was thus made to perform the function of the bolometer in measurements of radiant energy.

[1] "Phil. Mag.," Vol. XX, p. 173 (1910); Vol. XXI, p. 404 (1911).

There is, however, one important difference which requires consideration. The stream of electrons absorbed by the strip constitutes a current flowing out of that arm of the Wheatstone's bridge mesh. This current of itself will cause a deflexion of the galvanometer previously balanced when the thermionic current was not flowing. The deflexion which thus arises will depend both on the magnitude of the thermionic current and on the point at which it is allowed to return to the Wheatstone's bridge circuit. There is one such point in each of the adjacent arms for which this deflexion is zero, whether the battery circuit of the bridge is closed or not. In order to eliminate this difficulty, therefore, all that is necessary is to provide the resistance of one of the adjacent arms with a sliding contact connected to the line through which the thermionic current is returned, and to adjust the position of the contact until no deflexion is caused by turning on the thermionic current when the main Wheatstone's bridge current is off. Some other corrections, the chief of which are for the effect of lack of saturation when the currents are unsaturated and for the Joule heating effect of the thermionic currents, together with a number of minor possible sources of error, are discussed in the original papers.

We have seen that what is required to be measured in these experiments is the quantity of heat, which we may denote by J, liberated per absorbed electron, when the electrons have not acquired kinetic energy from an applied electric field. As a matter of fact it is necessary to employ *some* potential difference in order to drive enough electrons from the osmium to the strip to produce measurable effects; so that J cannot be measured directly. The value of J can, however, readily be deduced by making experiments with different, but sufficiently small, values of the applied potential difference V. Let V be expressed in volts and J in equivalent volts; that is to say, let J be the number of volts through which an electron would have to fall in order to acquire an amount of kinetic energy equivalent to the quantity of heat which it is desired to determine. Clearly, when the potential difference is V, the heat developed per electron, or per unit thermionic current, will be proportional to J + V; so that in order to determine J all we have to do is to divide the observed deflexions, for various values of V, by the corresponding thermionic currents and plot the resulting numbers against V. A linear relationship should thus be exhibited, and the value of J in volts should be equal to the intercept on the voltage axis between the point of intersection of the line through the experimental points and the position of zero volts. If the effect exists this intersection will be on the negative side of the zero. That these requirements are satisfied is shown by Fig. 20 which represents the results obtained with platinum. The points obtained with a platinum strip previously saturated with oxygen are shown thus: $\oplus$, with a platinum strip saturated with hydrogen thus: $\odot$, and for a fine platinum wire of circular section thus: $\bullet$. The points which fall on the two lines towards the right-hand side have been arbitrarily moved a distance corresponding to 4 volts in this direction to avoid confusion. It will be seen that the magnitude of J is not far from 6 volts and is slightly less for the strip soaked in hydrogen than for the others.

It is now desirable to consider the interpretation of J a little more carefully than we have done. Imagine a slab of the hot metal A at temperature T connected by a wire of the same material to a parallel slab of the cold metal B at temperature T_0, the whole being in a suitable vacuous enclosure. This system is not in equilibrium on account of the difference of temperature $T - T_0$, but may be considered to be artificially maintained in the definite condition just described. The work done in taking a single electron from A to B along a path passing across the space intervening between the metals may be denoted by

$$w_2 - e(V_2 - V_1) - w_1, \qquad . \qquad . \qquad (62)$$

where w_2 is the work at the surface of A, w_1 that at the surface of B, and V_2 and V_1 are the potentials at points just outside A and B respectively. The work between the same points for a path along the wire and never passing outside the metals may be written

$$\epsilon \left\{ P_{12} + \int_{T}^{T_0} \sigma_2 dT \right\} \qquad . \qquad . \qquad . \quad (63)$$

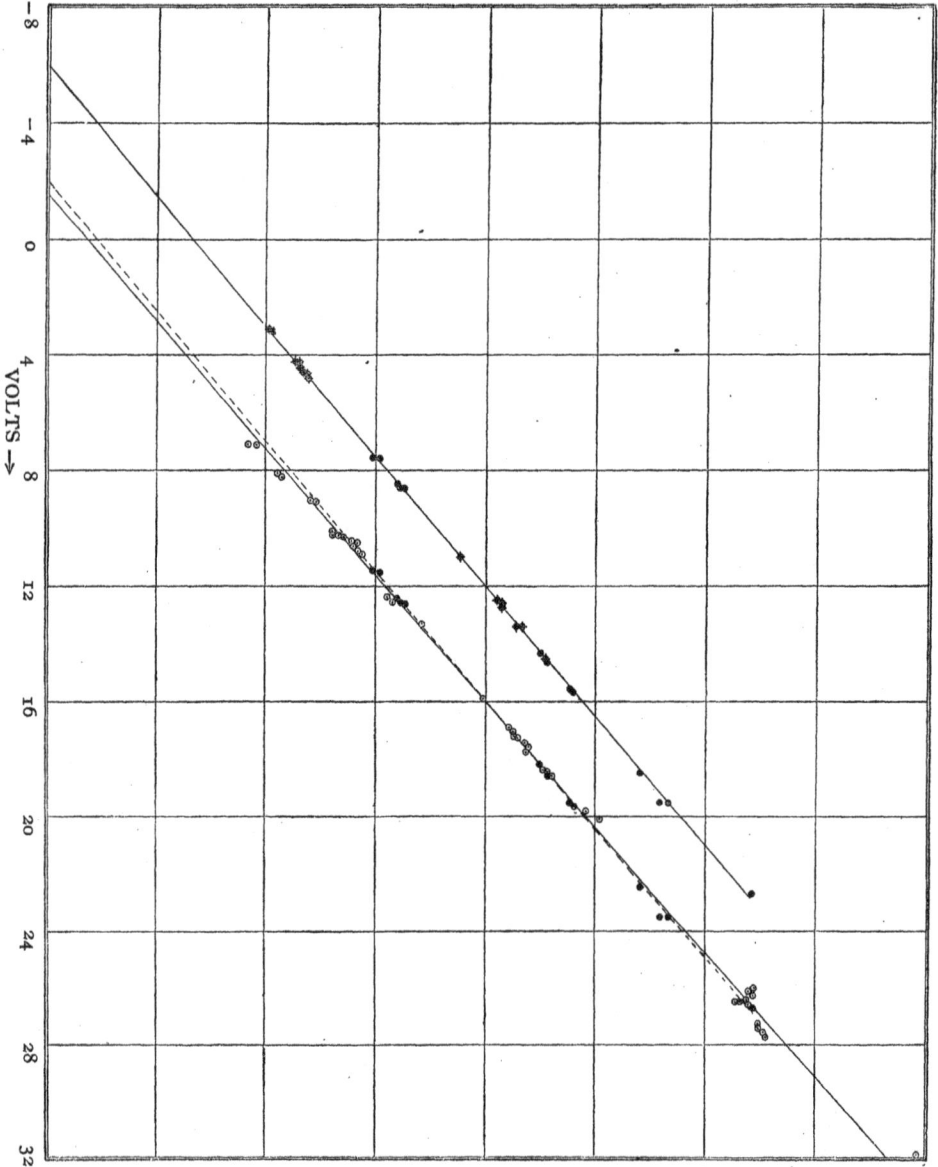

FIG. 20.

where P_{12} is a quantity comparable with the Peltier effect at the junction and σ_2 is comparable with the specific heat of electricity in A. The difference between the symbols and the corresponding thermoelectric quantities depends on the conditions which are supposed to govern the behaviour of electrons in metals; but there is no reason to suppose that the quantities are not of the same order of magnitude. The expressions (62) and (63) must be equal; so that

$$w_1 + \epsilon(V_2 - V_1) = w_2 + \epsilon\left\{\int_{T_0}^{T} \sigma_2 dT - P_{12}\right\} \quad . \quad (64)$$

The second term on the right is small compared with w_2; so that $w_2 - w_1$ has nearly the same value as $\epsilon(V_2 - V_1)$.

Now consider the heat liberated as each electron passes from A to B under the conditions of the experiments, when, however, the externally applied potential difference V is zero. First suppose that w_2 is greater than w_1. A is then electronegative to B and $V_2 - V_1$ accelerates electrons going from A to B. The average kinetic energy of the electrons when they leave A is, as we saw at the beginning of this chapter, equal to $2kT$. At a point immediately outside of B it has increased to $2kT + e(V_2 - V_1)$ and after passing through the surface it becomes $2kT + e(V_2 - V_1) + w_1$. If K_1 is the kinetic energy of the internal electrons of B at temperature T_0, the quantity of heat J liberated by the electron considered will thus be given by

$$J = 2kT + e(V_2 - V_1) + w_1 - K_1$$
$$= 2kT + w_2 - K_1 + eP_{12} - e\int^{T} \sigma_2 dT$$

On certain views of the behaviour of electrons in metals the last three terms are equal to $-K_2$ where K_2 is the kinetic energy of the internal electrons in A. In any event they are not likely to differ seriously from this quantity; so that to a sufficient degree of approximation

$$J = 2kT + w_2 - K_2 = 2kT + \Phi_0{}^2, \quad . \quad (65)$$

where $\Phi_0{}^2$ denotes the value of the quantity called Φ_0 on p. 185 for the substance A.

13

Now consider the case when $w_2 < w_1$ and A is electropositive to B. $V_2 - V_1$ will now retard electrons passing from A to B. This will cause the electrons reaching B from A with zero potential difference applied externally (V = o) to arrive at B with the same average kinetic energy as those which left A had, the only effect of the field being to reduce the number.[1] Their kinetic energy after passing the surface layer will thus be $2kT + w_1$ and the heat liberated

$$J' = 2kT + w_1 - K_1 = 2kT + \Phi_0^1 \qquad . \qquad (66)$$

where Φ_0^1 is the value of Φ_0 for the substance B.

It thus appears that if the hot metal A is electropositive to the cold metal B, the magnitude of J is determined by that of Φ_0 for the cold metal B. In this case the value of Φ_0 is greater for B than for A. If, on the other hand, the hot metal A is electronegative to B, the magnitude of J is determined by that of Φ_0 for the hot metal A. In this case, however, the value of Φ_0 is less for B than for A. Thus the heat liberated at the cold metal tends to be as great as possible, in accordance with the principle of increase of entropy, and the value of Φ_0 deduced from these measurements is the value belonging to that metal of the pair for which this quantity is greatest, quite apart from whether the metal functions as the receiver or emitter of electrons.

The values of Φ_0 calculated from the values of "ϕ" in the papers referred to are collected in the following table:—

Metal.	Actual Values of Φ_0.	Mean Φ_0.	b.
Platinum (strip) in O_2	6·31, 5·20, 5·47, 5·43, 5·35, 5·34	5·40	$6·65 \times 10^4$
Platinum (wire) in O_2 .	5·74, 5·47, 6·21, 5·21	5·66	$6·86 \times 10^4$
Platinum (strip) in H_2 .	{ 5·04, 4·10, 4·35, 4·48, 4·84, 4·21, 3·73, 5·38, 5·64, 5·60 }	4·70	$5·74 \times 10^4$
Gold . . .	6·39, 7·01, 6·68, 7·89, 7·21, 6·01	6·86	—
Nickel . .	5·04, 5·23, 5·46	5·24	—
Copper . .	7·21, 7·01, 6·71, 6·71, 5·61, 6·86	6·69	—
Phosphor bronze .	5·67	5·67	—
Palladium . . .	5·89, 5·29	5·57	—
Silver . . .	4·91, 5·68, 4·01, 5·31, 5·11	5·00	—
Aluminium . .	7·45, 6·75, 5·15, 7·75, 8·45	7·25	—
Iron . . .	{ 4·80, 4·03, 4·17, 4·70, 5·39, 5·39, 6·12, 6·43, 6·34, 7·37 }	5·49	—

The values of Φ_0 are in equivalent volts and the values of b in degrees centigrade. Unfortunately, the experimental

[1] Cf. O. W. Richardson, "Phil. Mag.," Vol. XVIII, p. 697 (1909).

numbers show a good deal of variation for each individual substance. Since the publication of those numbers the authors have spent a good deal of energy trying to improve the technique of the measurements, so as to get more accurate results, without any apparent success. The strips become discoloured during the experiments and the difficulties may be due to changes in the radiating power of the surface in the course of an experiment, and also to the discharge wandering from one part of the strip to another. It is possible that better results might be obtained by using tungsten as a source of electrons, as it is less volatile than osmium. As the results stand it is questionable whether much weight can be attached to the differences of Φ_0 shown in the last column but one; so that it may be that the values would show very little difference if they could be determined more accurately. This is to be expected on the theory outlined above, since all the metals below gold are probably electropositive to osmium and should give the value of Φ_0, characteristic of that metal. The values for gold and platinum also are not likely to differ much from that for osmium. Thus the numbers support the theory so far as they can be relied on. At least it is satisfactory to note that the values of Φ_0 are of the expected magnitude; and that the calculated values of b for platinum, for which metal the present data are much the most consistent and reliable, agree with the best values of that quantity as deduced from the temperature variation of the emission and given in the table on p. 81.

CHAPTER VI.

THE EMISSION OF POSITIVE IONS BY HOT METALS.

THE older experiments referred to in Chapter I showed that positive ions are liberated by hot metals under certain conditions. Thus, in 1873, Guthrie found that an iron ball in air at atmospheric pressure allowed positive but not negative electricity to leak away from its surface at a dull red heat. This experiment shows that positive ions are liberated at the surface of the metal but that negative ions are not; it does not show whether the positive ions arise from the interaction of the metal and the air, or whether they result merely from the high temperature of the metal. To determine this question it is necessary to make similar experiments in a vacuum. Such experiments, using electrically heated wires, were made by Elster and Geitel, who found that freshly heated metals, in general, emitted only positive ions when the temperature was not too high, and that the effect occurred both in a vacuum and in an atmosphere of various gases. At higher temperatures an emission of negative ions of the kind already considered accompanies this emission of positive ions; so that an insulated metal then discharges electrification of either sign.

These and many other experiments have abundantly shown that there is an emission of positive ions from freshly heated metals in a vacuum which has nothing directly to do with the presence of a surrounding gaseous atmosphere. The bearing of occluded gases on this emission is a different question which will be considered later (p. 222). The present chapter will be devoted to the conditions affecting this emission and the properties of the ions liberated thereby. One of the important features of the phenomenon, discovered by Elster and Geitel, is its transient character. If a metal is heated *in vacuo* at constant temperature, the positive emission is greatest at first and diminishes with time to smaller and smaller values. In this respect its behaviour affords a strong contrast to that of the negative emission. So far as the writer has observed there is no limit to the decay of the positive emission in a good vacuum ; so that the property is one which characterizes some exceptional condition of metals which have not previously been heated *in vacuo*, and is not a characteristic of the metal as such. It is possible that there are small positive emissions of a lower order of magnitude which do not decay with time, and which are definitely characteristic of the metals themselves, but there is no convincing evidence that they have yet been discovered.

In addition to the effects immediately under discussion there is an emission of positive ions from metals heated in atmospheres of various gases which is probably of different origin and which, at any rate, is much more permanent in character. This will be considered in Chapter VII.

The Decay of the Emission with Time.

The variation with time of the thermionic current from a positively charged platinum wire in a vacuum at a constant temperature has been examined by the writer.[1] The applied potential difference was constant and sufficient to produce saturation in the later stages of the experiments (see p. 200). The precise form of the current time curves varies from one specimen of wire to another. It also depends on the treatment of the wire and the temperature during the experiment. Fig. 21 exhibits some of the characteristic features. The current i decays rapidly at first and then more slowly, apparently approaching a constant value i_0 asymptotically. The curve shown can be represented by the equation

$$i - i_0 = Ae^{-k} \quad . \qquad . \qquad . \qquad . \quad (1)$$

when t is the time, and A and k are constants. This formula can be deduced from the assumption that the ions which carry the part $i - i_0$ of the current are produced by the decomposition

[1] " Phil. Mag.," Vol. VI, p. 80 (1903).

of some substance present in the wire, if it is admitted that
the rate of decomposition is proportional to the amount
of the substance present. This interpretation is a possible,
though not a necessary one. Similar results might follow if
the active substance were disappearing through evaporation or
diffusion. More complete experiments have shown that the
part i_0 of the current is not constant. It also decays with

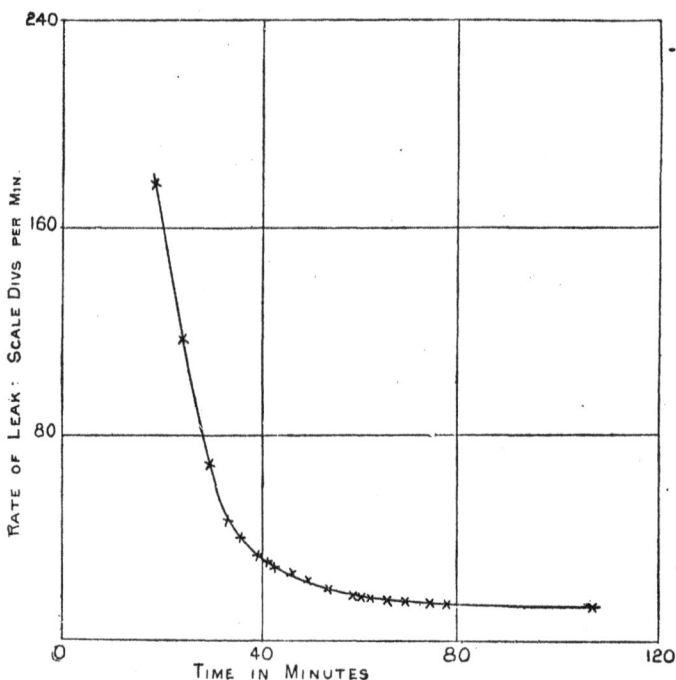

FIG. 21.

time in the same general way as the initial part, but much
more slowly.

Often the time changes are more complex than those
shown in Fig. 21, the quick initial drop being followed by a
later rise to a maximum, after which the emission shows the
final slow decay. An example of this type of change, taken
from a paper by the writer,[1] is shown in Fig. 22. These
effects have a superficial resemblance to radio-active changes
and may be interpreted in a somewhat analogous manner.

[1] "Congrès de Radiologie," Liège, C.R., p. 50 (1905).

It is probable that there are at least two substances concerned. One of these A may be supposed to decompose, emitting the ions which cause the large initial current which decays quickly, whilst a second substance B decomposes with little or no emission, but forms a third substance C which is relatively more active. It is not supposed that these changes have anything more than a formal analogy with radio-active processes ; it is probable that they are of a chemical character and that the emissions are characteristic of the various chemical pro-

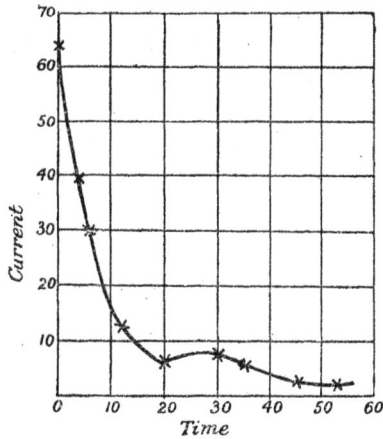

Fig. 22.

ducts. The conditions under which the intermediate hump appears have been insufficiently studied. So far as the writer's observations go their presence is favoured by relatively low temperatures. Similar effects have been observed by Sheard[2] at atmospheric pressure (p. 229).

CURRENT AND ELECTROMOTIVE FORCE.

The rate of decay of the initial emission just referred to increases rapidly with rising temperature. It is very small at the temperatures at which the emission is conveniently measurable with a sensitive electrometer. Experiments made at such relatively low temperatures enable the dependence of the thermionic current on conditions such as the magnitude of the applied electromotive force to be investigated, without having to consider complications arising from independent changes of the emission with the time.

The experiments which have been made to measure the current, with different applied potentials, from a positively charged hot wire to a suitable electrode in a good vacuum

[1] Liège, C.R., loc. cit.
[2] "Phil. Mag.," Vol. XXVIII, p. 170 (1914).

under these conditions have shown that the relation between current and potential difference is surprisingly complicated. The first observations, made by the writer,[1] indicated that the current was proportional to the voltage from + 40 to + 400 volts. This result is very surprising because the currents are very small, there are no negative ions emitted by the hot wire, and the positive ions, being of atomic dimensions, are so massive that their motion is unaffected by the presence of the magnetic field due to the current used to heat the wire. Thus all the conditions, namely : spatial density of the ions, recombination, and deflexion by the magnetic field, which may in general operate to prevent saturation, are absent. We should, in fact, expect these currents to be saturated by the application of any positive potential sufficient to make the negative end of the hot wire positive to the collecting electrode after allowing for the drop due to the heating current.

The phenomena usually observed with voltages under 40 are, to a certain extent, more in line with expectation. Thus the writer[2] found that with a new wire in air at 0·0015 mm. the current increased as the potential was raised from 0 to about 3 volts where it showed signs of saturation. On increasing the potential from 3 to 40 volts there was a steady *decrease* in the value of the current. Thus under certain circumstances the currents may diminish with rising voltage. This statement refers only to the relatively steady values which are obtained after the potentials have been applied for a few minutes. The initial currents are usually larger if the potential has been raised, and sometimes smaller if it has been lowered, immediately before the observation.

Further experiments on the subject have been made by the writer and C. Sheard.[3] The current from a hot platinum wire at various voltages was measured in three different types of apparatus at pressures recorded by the McLeod gauge as

[1] " Phil. Mag.," Vol. VI, p. 80 (1903).

[2] " Phil. Trans., A.," Vol. CCVII, p. 11 (1906).

[3] " Phys. Rev.," Vol. XXXIV, p. 391 (1912); " Phil. Mag.," Vol. XXXI p. 497 (1916).

under 0·0002 mm. With new wires the current grew to a maximum as the potential was increased from 0 to something below 5 volts; it then usually diminished a little, and finally increased, being roughly proportional to the potential from + 40 to + 400 volts. So far these results confirm those already described. In some cases, however, the drop in the current after passing 5 volts was not observed. The increase from 40 to 400 volts was found gradually to die away as the heating continued. The writers concluded at the time that this part of the current was due to the bombardment of the wire by electrons liberated by the impact of the positive ions on a layer of gas at the surface of the negative electrode. It is questionable, however, whether this interpretation can be considered to be established definitely without further experiments. It appeared that after the increased current at high potentials had been destroyed by continued heating it could be restored, to a greater or less extent, by the following agencies: (1) heating the positively charged wire to a higher temperature than any previously employed; (2) allowing a discharge of negative electrons to pass from the hot wire to the cold electrode, and (3) admitting air to the apparatus. However, later experiments by H. H. Lester[1] indicate that the effectiveness of these agents is not always to be relied on. The observed effects, in fact, may be due not to the causes mentioned but to some unknown factor which was altered at the same time.

It is evident that the drop in the current sometimes observed when the potential is increased beyond 5 volts, and the large increase above 40 volts, still require explanation. These phenomena are shown only by new wires when first heated at a low temperature; but similar or related effects are exhibited in a gaseous atmosphere as well as at the lowest pressures (see p. 231). The experimental investigation of these effects is extraordinarily difficult as it is very hard to reproduce the same conditions in successive experiments. Similar effects, but usually not so well marked, are often

---

[1] "Phil. Mag.," Vol. XXXI, p. 549 (1916).

exhibited by the negative (electronic) emission from freshly heated wires. The difficulty of attaining saturation of the electron currents which is peculiar to new wires has already been referred to (p. 66).

REVIVAL OF OLD WIRES.

A wire which has lost the power of emitting positive ions through continued heating in a vacuum can be revived in a number of ways, some of which give important indications as to the cause of the emission. The various methods will be considered in order.

1. By distillation. The writer[1] found that if an old wire A was mounted near a fresh wire B, and B was heated and charged positively, A being cold, the passage of the thermionic current from B to A caused A to re-acquire the power of emitting positive ions when heated again. The same thing occurred, but to a smaller extent, if B was negative with respect to A or if they were at the same potential. These experiments indicate that the emission is, at least in part, due to a substance which may be distilled from one metal to another. The fact that the effect is greatest when the wire B is positively charged indicates that the ions emitted by B are either themselves re-emitted or cause the formation of new ions when A is heated afterwards.

2. Effect of a luminous discharge.[2] The power of emitting positive ions on subsequent heating is restored if an old wire is placed in a tube through which a luminous discharge is caused to pass in various gases at a low pressure. The effect is greatest if the wire is close to the cathode and is inappreciable at distances exceeding a few centimetres. It also disappears if the wire is shielded from a direct view of the cathode by a solid obstacle, indicating that the revival is caused by something projected from the cathode. However, this seems to be only part of the story because separate experiments showed the wire was revived when it was itself made the cathode during the passage of the dis-

---

[1] "Phil. Mag.," Vol. VI, p. 86 (1903).
[2] O. W. Richardson, Vol. VIII, p. 400 (1904).

charge. It seems likely that the sputtering of the surface of an old wire under these conditions exposes fresh material which has not lost the power of positive emission.

The reviving effect produced by an auxiliary cathode occurred with cathodes of platinum, aluminium, and carbon when the gas was either air, oxygen, or nitrogen. The effects in hydrogen were very small. At moderate pressures the effect increased as the pressure diminished; thus in air when the pressure was reduced from 0·8 to 0·0025 mm., and the discharge passed for a given time, the quantity of electricity emitted when the wire was subsequently heated was increased by a factor of about 300. By a kind of fractional distillation of the imparted emissibility it was possible to show that the effects were due to the formation on the wire of two distinct substances.

The bulk of the observations which have been made with this effect indicate that it is intimately connected with the sputtering of metal from the surface of the cathode. On the other hand, similar experiments made by Garrett[1] on the effect of a discharge in carbon dioxide on the emission of positive ions from aluminium phosphate (see p. 270) led him to conclude that the revival occurred only when fresh gas had been admitted to the apparatus.

3. The writer[2] observed that the emission from an old wire was enormously increased if the walls of the glass tube in which it was mounted were slightly heated. This effect occurs if the glass and platinum are carefully cleaned with acid and dried before testing. In a particular experiment it was found that warming the glass walls with a Bunsen burner for about two minutes increased the current from the positively charged hot wire from $2 \cdot 2 \times 10^{-13}$ amp. to 5×10^{-9} amp. The effect is not caused by ordinary gases expelled from the glass by heating, as the pressure rose only from 0·0005 to 0·001 mm. in this experiment, and the positive emission caused by any of the commoner gases at these pressures is negligible in comparison with the observed currents.

---

[1] " Phil. Mag.," Vol. XX, p. 572 (1910).
[2] " Phil. Trans., A.," Vol. CCVII, p. 19 (1906).

4. Exposure to gases at high pressures. Klemensiewicz[1] found that an old wire is revived by exposure to atmospheres of hydrogen, nitrogen, or oxygen at pressures of 50 to 100 atmospheres at a temperature in the neighbourhood of 200 C. He concludes that the initial ionization from fresh wires is therefore due to absorbed gases (see p. 222).

5. Heating in a gaseous atmosphere. Various observers have recorded that old wires are revived when heated for a short time in an atmosphere of various gases or in a Bunsen flame.

6. Straining. The writer[2] found that a manganin wire was revived when subjected to the strain caused by passing a current through it in a varying magnetic field.

The processes just described all give rise only to effects of a somewhat temporary character. The increased emission rapidly disappears when the exciting agency is no longer operative and the wire is subsequently heated in a vacuum. In addition to the effects enumerated, an old wire may exhibit an increased emission of a comparatively permanent character when it is immersed in a gaseous atmosphere. The effects which then arise will be considered fully in Chapter VII.

It will be seen from the foregoing list that almost any change which may be made in the condition of an old wire restores, to some extent, its power of emitting positive electricity. There are, however, two processes which might conceivably be expected to produce such an effect and which do not do so. An old wire is not revived either by exposure when cold to dust-free air at atmospheric pressure or by being allowed to stand in the cold for long periods of time in a vacuum. At one time it was thought that the first of these agencies did produce an effect,[3] but further investigation[3] showed that it was due to other causes. Experiments[4] made to test the second point have only extended over a period of three months, but there is no definite reason for expecting

[1] " Ann. der Physik," Vol. XXXVI, p. 796 (1911).
[2] " Roy. Soc. Proc., A.," Vol. LXXXIX, p. 521 (1914).
[3] O. W. Richardson, " Phil. Mag.," Vol. VI, p. 90 (1903) ; Vol. VIII, p. 410 (1904).
[4] O. W. Richardson, " Congrès de Radiologie," Liège, C.R., p. 53 (1905).

that longer intervals would be much more likely to lead to a positive result.

VARIATION OF EMISSION WITH TEMPERATURE.

We have seen already that at very low temperatures the rate of decay is so small that the initial positive emission can be regarded as a function of the temperature. The first experiments to measure the positive emission at different low temperatures were made by Strutt.[1] The currents were measured with an electroscope and the electrostatic capacity of the system was quite small. Thus by taking the deflexions over rather long intervals of time very small currents could be measured. The following emissions were investigated: copper in air, copper oxide in hydrogen, silver in air, silver in hydrogen, and copper oxide in air. In each case the pressure of the gas was 1·0 cm. The following values given by a silver wire in air may be considered typical, as there is no striking difference in the results given by the different materials examined :—

Temperature °C.	194	210	217	227	240	258
Current	0·2	0·84	1·46	5·0	6·2	45·6

If the capacity, which is not given, is taken to be 10 cm. the unit of current would be about 3×10^{-15} amp. In each case measurable effects were obtained in the neighbourhood of 200° C. and the currents increased very rapidly with rising temperature. The temperature at which a current of 10 units was obtained was lower with silver wires in air and hydrogen, and with an oxidized copper wire in air, than with a clean copper wire in air or an oxidized copper wire in hydrogen, indicating that chemical action is unfavourable to the emission rather than the reverse.

The writer[2] pointed out that the currents observed by Strutt followed the formula $i = AT^{\frac{1}{2}}e^{-b/T}$, with A and b constants, which, as we have seen, governs the variation of the negative emission with temperature. It was also pointed out

---

[1] " Phil. Mag.," Vol. IV, p. 98 (1902).
[2] " B.A. Reports," Cambridge, 1904, p. 473.

that the same formula also covered the following cases : (1) the positive emission from a wire revived by the electric discharge, when measured at temperatures such that the time-rate of decay is small (p. 181) ; (2) the positive emission from the alkaline earth oxides heated on platinum at atmospheric pressure as observed by Wehnelt,[1] and (3) Owen's[2] observations on the positive emission from the Nernst filament at low pressures. In fact the formula has been found to cover all cases of emission of both positive and negative ions from solids in which the emission can be considered to be a definite function of temperature.

The value of b for the initial positive emission is much smaller than that of the corresponding constant when the formula is applied to the emission of electrons. Thus Strutt's numbers for silver in air give $b = 1\cdot34 \times 10^4$ degrees C. This is less than one-fourth of the value of the corresponding quantity for the electronic emissions for most of the elements given in the table on p. 81. Thus, if they could be compared at the same temperature, the negative emission would be found to increase much more rapidly than the positive ; so that, apart from the decay of the positive emission with continued heating, there is an additional reason why the positive emission should inevitably be swamped by the negative at high temperatures.

THE KINETIC ENERGY OF THE IONS.

The distribution of kinetic energy among the positive ions emitted from a heated platinum strip was first examined by the writer,[3] using the methods described on p. 167 in connexion with the same problem for the energy of the negative electrons. It was found that the results were in agreement with the view that the distribution of velocity among the positive ions was in accordance with Maxwell's Law and that their average kinetic energy was the same as that of the mole-

---

[1] " Ann. der Physik," Vol. XIV, p. 425 (1904).
[2] " Phil. Mag.," Vol. VIII, p. 249 (1904).
[3] *Ibid.*, Vol. XVI, p. 890 (1908).

cules of a gas at the same temperature as the hot metal. As in dealing with the negative electrons, this conclusion was established both by a consideration of the comparative magnitudes of the current through the slit at different distances from the central position, by the actual magnitude of the current in the central position through a slit of given width, and by the inferred value of the gas constant R deducible from the experimental results on the assumption that the ionic charge is the same as that of a monovalent ion in electrolysis. The three methods of determining R referred to on p. 167 led to the respective values $4\cdot0 \times 10^3$, $3\cdot3 \times 10^3$, and $5\cdot4 \times 10^3$. The mean of these is $4\cdot2 \times 10^3$ as against the theoretical value $3\cdot7 \times 10^3$. When all the possible sources of error and the limitations of the apparatus used are taken into consideration this agreement is as good as could reasonably be expected.

The experiments just described supply us with information about that part only of the kinetic energy which depends on the component of velocity of the ions parallel to the emitting surface. The investigation was extended to include the normal component of the velocity by F. C. Brown,[1] using the method described on p. 156. The current i between parallel plates against an opposing potential difference V was found, as with the negative electrons, to satisfy the equation

$$\log \frac{i}{i_0} = -\frac{\nu\epsilon}{RT}V \qquad . \qquad . \qquad . \qquad (2)$$

required by Maxwell's Law and deduced on p. 159. The values of R varied from $3\cdot5 \times 10^3$ to $4\cdot0 \times 10^3$, the mean being $3\cdot6 \times 10^3$ instead of $3\cdot7 \times 10^3$. Brown also found that the results of the experiments were independent of the pressure of the surrounding gas between the limits 0·009 mm. and 28·0 mm. These experiments were limited to platinum, but in a later paper Brown[2] extended the observations to cover a large number of substances, using the same general method. Where possible the materials tested were in the form of discs or strip, but in some cases thin wires or filaments had to be used.

[1] " Phil. Mag.," Vol. XVII, p. 355 (1909).
[2] *Ibid.*, Vol. XVIII, p. 649 (1909).

Generally speaking, wires or filaments were found to give values of R higher than the normal, probably owing to distortion of the electric field near the hot metal. An osmium filament, however, gave an abnormally low value of R ($2 \cdot 5 \times 10^3$), but this was attributed by Brown to the presence of electrons, which modify the conditions affecting the motion of the ions. The final values of R and the data leading up to them are collected in the following table :—

Material Tested.	Form of Emitter.	Absolute Temperature.	Current i_0 (when V = 0) $I = 10^{-12}$ amp.	Pressure (mm.).	$R \times 10^{-3}$.
Gold I . .	disc	$\left\{\begin{matrix}1030\\973\end{matrix}\right\}$	1·0	0·007	4·2
Gold II . .	,,	$\left\{\begin{matrix}1190\\1163\end{matrix}\right\}$	60·0	0·01	3·9
Silver I . .	,,	1020	0·8	0·002	3·0
Silver II .	,,	1150	35·0	0·008	2·9
Palladium . .	,,	1170	25·0	0·04	3·4
Nickel . .	strip	1120	2·5	0·003	3·6
Iron I . .	disc	1100	1·0	—	4·6
Iron II . .	,,	1100	0·8	0·005	5·2
Iron III . .	,,	1240	—	0·01	4·4
Platinum I. .	wire	1695	—	—	5·1
Platinum II .	disc	1293	5·0	0·009	3·5
Tungsten . .	filament	1150	1·0	0·0003	5·1
Tantalum I .	,,	1050	0·7	0·0005	9·6
Tantalum II .	strip	1050	1·0	0·002	3·0
Osmium . .	filament	1120	3·0	—	2·5
Aluminium phosphate I . .	disc	1230	100·0	—	3·9
Aluminium phosphate II . .	,,	1170	120·0	0·006	3·4

If the values with wires and filaments are disregarded as not fulfilling the conditions laid down by the theory of the experiments, the average of the remaining numbers in the last column is $R \times 10^{-3} = 3 \cdot 8$, instead of $3 \cdot 7$. The values for a given metal with a given form of radiator agree better with one another than do the values when different metals are compared. Thus all the values for iron are distinctly high and for silver distinctly low. The difference between iron and silver is greater than could be expected from any obvious source of experimental error, and Brown concludes that there is a real difference in the value of R for the ions from different metals. It is questionable, however, whether this inference can be accepted without further experimental support. The difference between the results for a strip and a filament of tantalum is much greater than the difference given by discs of iron and silver, showing that relatively small differences in the geometrical arrangement may make very great differences in the final values. There is also the difficulty arising from the presence of negative electrons in some instances, a factor which is very difficult either to control or allow for. The temperature, too, is difficult to determine with an apparatus of the type used. Taking all the circumstances into account it would appear that the only inference which can be drawn with certainty from these experiments is that the distribution of energy is not far from that required by Maxwell's Law in the case of the positive ions from all the substances examined.

It is necessary to add that Schottky[1] has announced that he has obtained evidence of much greater values for the energy of the positive ions emitted by hot bodies than those given by the foregoing investigations. As no details are given it is impossible to state what is the probable cause of this discrepancy.

THE CHARGE OF THE IONS.

If we regard it as inherently probable that the distribution of velocity among the emitted ions is in accordance with Maxwell's Law, the foregoing experiments may be used to demonstrate that the charge of the positive ions is equal to that of an electron or of a monovalent ion in electrolysis. In some of the experiments with platinum strips which had been heated for a long time, the writer[2] found that the spreading out of the ions was abnormally small. The results could be reconciled with either of the two following hypotheses: (1) that the charge was equal to e but that the kinetic energy had only half the normal value, and (2) that the kinetic energy had the normal value but that the charge was doubled. Of the two alternatives the latter is more likely to be true. At the same time enough experiments have not been made on this particular subject to make it certain that the observed effect is a real one. Bending outwards of the strips tested or the appearance of electrons, both of which tend to occur

[1] " Ann. der Physik," Vol. XLIV, p. 1030 (1914).
[2] " Phil. Mag.," Vol. XVI, pp. 900, 903, 906 (1908).

14

with continued heating, would produce effects in the direction of those observed and might be capable of accounting for the phenomena. Until more detailed experiments are forth-coming it is not desirable to attach too much weight to this particular piece of evidence of the occurrence of ions with multiple charges in the positive emission from hot metals. Another point which tends to make this evidence doubtful is that the value of ϵ/m for these particular ions, which was measured at the same time as the energy, would make their electric atomic weight equal to about 100, a value which would be difficult to harmonize with that for any substance likely to be present.

A direct attempt to measure the charge of the ions from hot bodies at atmospheric pressure has been made by Pomeroy.[1] He concluded that the positive ions emitted by a platinum strip had a charge 2ϵ on first heating, but that the average value of the charge gradually fell to ϵ as the heating was continued. The method used was one due originally to Townsend, but in applying it the importance of the mutual repulsion of the ions has been insufficiently considered,[2] and it is doubtful what interpretation ought to be put on the results obtained. In any event, the conditions affecting the origin and motion of the positive ions in these experiments are quite different from those present when the kinetic energy and specific charge (ϵ/m) have been measured.

THE SPECIFIC CHARGE AND ELECTRIC ATOMIC WEIGHT OF THE IONS.

In discussing the specific charge (ϵ/m) for the positive ions it is convenient to introduce a related quantity which we may call the electric atomic weight (M) of the ions. The last-named quantity is obtained when we divide the specific charge of a monovalent element of unit atomic weight, the value of which is 9649 E.M. units, by the specific charge of the ions under consideration. If the charge of a positive ion

---

[1] "Phil. Mag.," Vol. XXIII, p. 173 (1912).
[2] Townsend, *ibid.*, Vol. XXIII, p. 677 (1912).

is equal to that of an electron, and there are numerous indications that such is usually the case, the electric atomic weight is,
if we neglect the mass of the electron compared with that of
the atoms, equal to the chemical atomic or molecular weight of
the ions. In any event, even if the charge is a multiple of ϵ,
the electric atomic weight is, to the same close approximation,
equal to the chemical equivalent weight (in terms of $O = 16$)
of the same ion in electrolysis. Thus the determination of
ϵ/m or M is of the utmost importance if we wish to discover
the structure of the ions in question.

The first experiments to measure these quantities for the
positive ions from hot bodies were made by Sir J. J. Thomson[1]
who used the method of crossed electric and magnetic fields
described on p. 8. If d is the distance between the wire and
the receiving plate, H the magnetic intensity, and V the difference of potential just necessary to drag the ions across in
the presence of the magnetic field, then

$$\epsilon/m = 2\mathrm{V}/\mathrm{H}^2 d^2 . \qquad . \qquad . \qquad . \quad (3)$$

This formula applies[2] even if the electric field is not uniform,
a point not brought out in our earlier discussion. It assumes,
however, that the ions set out with negligible velocities, and
this is not quite correct. The experiments were made with
wires of iron and platinum which had been heated for a long
time in a vacuum. It was found that the behaviour of the
currents in a magnetic field was very capricious. In some
cases the thermionic current was unaffected by a magnetic field
of 19,000 gausses. When the currents were sensitive to the
magnetic field Thomson[3] describes the phenomena occurring as
follows : "When the potential difference between the hot metal
and the plate connected with the electrometer was small, say
3 or 4 volts, the leak was very nearly stopped by the magnetic
field ; with a potential difference of 10 volts the leak was reduced
by the magnetic field to about one-quarter of its original value,
the effect of the magnetic force upon the leak diminished as

---

[1] "Conduction of Electricity through Gases," 2nd edition, pp. 145, 217
(Cambridge, 1906).

[2] Thomson, loc. cit., p. 219.

[3] Loc. cit., p. 220.

the potential difference increased but was appreciable until this reached about 120 volts. Thus in this case we see that while some of the carriers can reach the plate under a difference of potential of 10 volts, there are others which require a potential difference of 120 volts to do so." With the dimensions of the apparatus used and with H = 19,000, for ions which are just stopped when V = 10, ϵ/m = 60, and M = 161, for those just stopped when V = 120, ϵ/m = 720, and M = 13·4. From these numbers Thomson concludes that the ions are a mixture of atoms of platinum and of the gas. The experiments were made in air, at 0·007 mm. pressure. If the atoms of the elements under consideration carried a charge ϵ the values of M would be 192 and 14 (or 16) respectively. Since more than half the current was stopped with 10 volts, the experiments indicate that the lighter ions carried the greater part of it. The carriers of the current in the condition in which it was un-affected by the magnetic field, under low potential differences·in the neighbourhood of 1 or 2 volts, are attributed to charged particles of platinum dust sputtered from the wire. Similar experiments made with iron wires gave ϵ/m = 400 and M = 24.

A fuller discussion of the interpretation of the numbers above will be given later (p. 223). Without going into detail it is clear that most of the ions under consideration are of atomic magnitude.

Measurements of ϵ/m and M for the positive ions from hot metals have been made by the writer[1] by a different method. The apparatus, which is similar to that used in investigating the distribution of the tangential component of emission velocity (p. 162), is shown diagrammatically in Fig. 23. The part on the left-hand side of the dotted line FF represents a section, by the plane of the paper, of two parallel metal plates AA and BB. The plates AA are fixed, and separated by a narrow slit of constant width running perpendicular to the plane of the figure. A narrow strip c of the metal to be tested is shown in

[1] "Phil. Mag.," Vol. XVI, p. 740 (1908); " Roy. Soc. Proc., A.," Vol. LXXXIX, p. 507 (1914); cf. also C. J. Davisson, " Phil. Mag.," Vol. XXIII, p. 121 (1912).

transverse section. It almost fills the slit and its upper surface is flush with that of the plates AA. The upper plates
BB also are divided by a narrow slit *a* parallel to the former.
Behind this is a box-shaped electrode indicated by E. The
plates BB and the electrode E are rigidly bolted together, BB
and E being insulated from one another. The rigid system
BBE can be moved backwards and forwards through small
measured distances in a horizontal line by the accurate screw
shown on the right. The parts ABE are all enclosed in a
glass tube permitting the attainment of a high vacuum. In

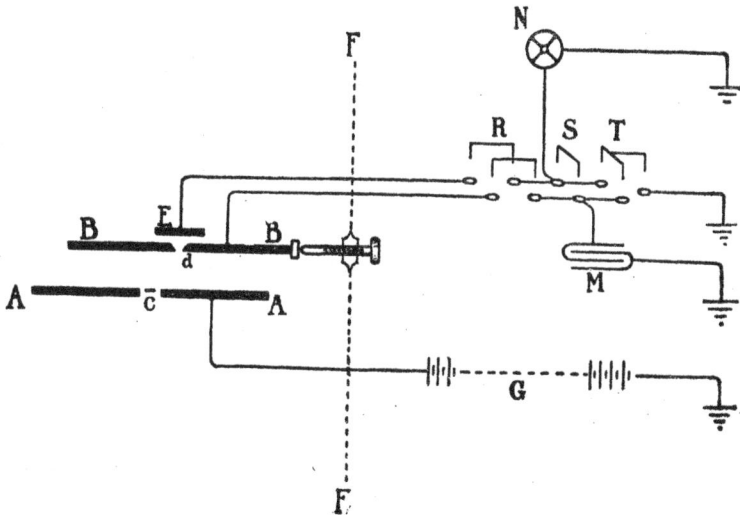

Fig. 23.

the final form of the apparatus the screw was provided with a
micrometer head and cyclometer attachment on the horizontal
axis. Both these were enclosed in the glass tube and read from
outside, the turning being effected by a right-angled bevel-
gear operated through a ground glass joint in a side tube.
The central region between the plates can be placed in a very
uniform magnetic field of measured strength running perpendicular to the plane of the paper. A suitable potential difference V is applied between AA and BB so as to drive the
ions on to the upper plate.

There are two ways of using the apparatus. These, for

convenience, will be called the slit method and the balance method respectively. The figure shows the electrical connexions for using the slit method. Initially the keys R and T are depressed; so that the electrometer N, capacity M, and BB and E are all connected to earth. On breaking T the charge passing through the slit into E flows into the electrometer, whilst that received by the plates BB flows into the capacity. When a suitable deflexion has accumulated, the compound key R is taken out, thus breaking both currents. The steady deflexion of the electrometer thus measures the quantity of electricity which has passed through the slit. The key S is then depressed; so that M and N are connected together. After the electrometer has come to rest the new steady deflexion will measure the quantity of electricity received both by the slit and the plates during the identical time interval in which the quantity previously measured passed through the slit. These measurements are repeated for different horizontal displacements x of the slit from the central position. The procedure by this method in fact is the same as that followed in getting the distribution of the tangential component of velocity, and described in Chapter V. The chief difference between the two experiments arises from the presence of the magnetic field. If the ions all have the same value of ϵ/m, and if the magnetic field is not too large, the only effect of its presence is to displace the resulting curve, which connects the proportion passing through the slit with x and is similar to Fig. 18, bodily to the left or to the right according to the direction of the magnetic intensity H. There is no distortion of the curves unless the ions are a mixture having different values of ϵ/m. In that case, if there is sufficient difference in the values of ϵ/m for the constituents, the curve which has a single maximum in the absence of a magnetic field develops two humps when the field is applied. Thus this method enables us to form a judgment as to the homogeneity of the ions. When the ions are homogeneous the value of ϵ/m is given by the formula

$$\frac{\epsilon}{m} = \frac{9}{2} \frac{V x^2}{H^2 z^4} \qquad . \qquad . \qquad . \qquad . \quad (4)$$

if x is the displacement of the maximum caused by the field H, and z is the distance between the plates. This formula is only an approximation, but it is correct to about 0·5 per cent under the conditions of the experiments.

The measurements when the balance method is used are much simpler. E is connected to one of the plates BB. These are insulated from each other and connected to the alternate pairs of quadrants of the electrometer, all four quadrants being insulated. Under these conditions the electrometer will not deflect if the potentials of each pair of quadrants change at equal rates. This happens when the dividing line between the upper plates is at the value of x corresponding to the maximum. If the screw is turned a little one way the electrometer deflects to the right, if in the other way to the left. The position of zero deflexion can be determined easily with precision, and two experiments with H in opposite directions are all that is required to determine the corresponding displacement $2x$ of the maximum required in equation (4). Thus this method enables e/m to be measured very rapidly. On the other hand, it does not tell anything about the homogeneity of the ions under investigation; so that it is advisable to restrict its application to cases in which the homogeneity of the ions has previously been established by experiments using the slit method, provided the latter method can be applied. The behaviour of the curves in a magnetic field indicates that as a general rule the ions which carry the large initial current from hot metals are very homogeneous. This conclusion is also supported by the fact that the value of e/m given by the measurements often remains constant over long periods of time (see below, p. 218).

The first experiments by the slit method, made partly in collaboration with E. R. Hulbert,[1] gave the values shown in the next table :—

Substance.	Value of e/m. (E.M. Units per Gm.)	Value of M (O = 16).
Platinum	361	26·8
Palladium	317	30·5
Copper	342	28·3
Silver	322	30·0

[1] " Phil. Mag.," Vol. XX, p. 545 (1910).

Substance.	Value of ϵ/m. (E.M. Units per Gm.)	Value of M (O = 16).
Nickel	357	27·1
Osmium	395	24·5
Gold	206 → 418	47 → 23·1
Iron	726 → 457	13·3 → 21·1
Tantalum	186 → 376	52 → 25·7
Tungsten	230	42·1
Brass	336	28·8
Steel	322	30·0
Nichrome	395	24·5
Carbon	332	29·1

With the apparatus as it was used in these experiments there are two sources of error of unknown magnitude, arising respectively from lack of uniformity of the electric field, and from the bowing of the strip due to its expansion. It was attempted to correct for these by using the same method and apparatus to measure the value of ϵ/m for the negative electrons from platinum. It is probable that this method overdoes the correction, since the electrons come off at a higher temperature than the positive ions; so that the corresponding errors would be greater in this measurement. The true values of ϵ/m are probably smaller, and of M greater, than those given in the table. The values have been corrected for an erroneous value of ϵ/m for the negative electrons which was used in the original papers.

The behaviour of tungsten was found to be erratic and the values given are relatively less reliable than the others. With gold, iron, and tantalum the numbers obtained on first heating were different from those obtained later. The later values persisted for a considerable time, until the emission disappeared or the material melted, in fact. No definite change of ϵ/m with time was noticed with the other substances. Excluding tungsten the relatively permanent values of M all lie between 21·1 and 30·5, the average being 26·9.

No great accuracy is claimed for the numbers found, on account of the reasons given above. At the same time a number of important inferences can be drawn from them. They show that the ions are not atoms or molecules of the elements concerned, since the values of M all lie between 20 and 30, whereas the atomic weights range from 12 for carbon to 192 for platinum. The similarity of the values indicates that the majority of the ions arise from some impurity common to all the metals. This impurity cannot be hydrogen or any light gas with an electric atomic weight below 20, as the values of M are too high for such bodies. The ions might be charged atoms of sodium or potassium or charged molecules of nitrogen, oxygen, or carbon monoxide or a mixture of these. To differentiate these various possibilities it was necessary to increase the accuracy of the experiments.

Substantial improvements in this respect have recently been effected by the writer.[1] Probably the most important source of error in the old experiments arose from the bowing of the strips when heated. This has now been eliminated by keeping them under a slight but sufficient tension. Since the distance z between the plates enters into the formula for ϵ/m through its fourth power, it is important that it should be capable of accurate and reliable measurement. This was accomplished by enlarging the apparatus and improving the technique in various ways. Another point to which insufficient attention had previously been paid was the accurate measurement of the magnetic field under the actual experimental conditions. Finally, any lack of uniformity in the electrostatic field was almost completely eliminated by reducing the gap between the edges of the strip and the sides of the slit almost to the vanishing point. The measurements were ultimately checked by comparing the values of M given by the metals tested with those given by the ions from potassium sulphate. These ions are known on independent grounds to be atoms of potassium which have lost an electron (Chapter VIII). The apparatus as improved has one disadvantage. It restricts the number of metals which can be experimented with. Most of those in the previous table were found either to yield or break under the tension required to keep the strips taut, at the temperatures at which the emission became copious enough conveniently to make measurements with.

Most of the measurements with the improved apparatus were made by the balance method, as this enabled any changes in ϵ/m with lapse of time to be followed readily. The two

[1] " Roy. Soc. Proc., A.," Vol. LXXXIX, p. 507 (1914).

methods, however, were compared in independent experiments and were found to give identical results when the emission was homogeneous, and was not changing in character with lapse of time. With platinum strips the ions given off during the first twenty hours or so of heating were found to be very homogeneous and to have a value of M very close to 40. In the later stages there were indications of ions for which M was near 23, and sometimes also values of M between 50 and 60 were obtained in the last stages of heating just before the strips broke. The variation of M with the time of heating is exhibited in Fig. 24, where the points marked thus

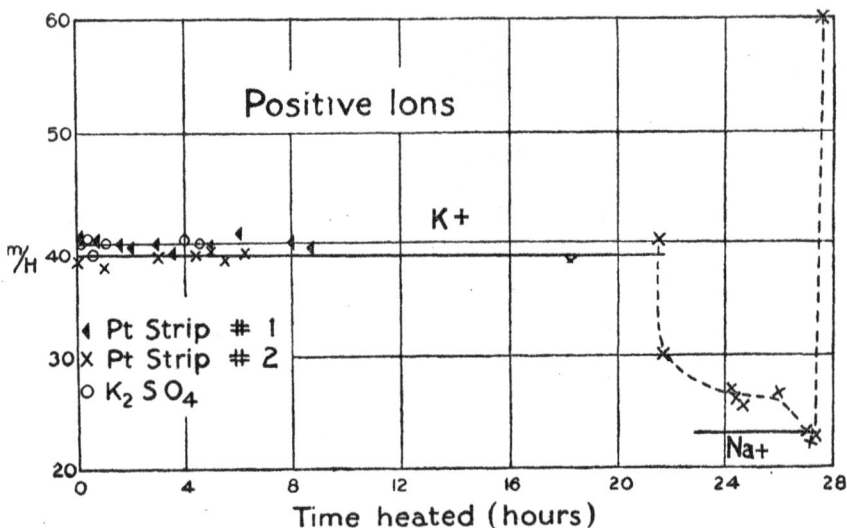

Fig. 24.

x represent some of the values given by a particular strip at times extending over twenty-eight hours. The final high values are a little uncertain as the strips never last long after these values have begun to appear. The points thus ◀ in the figure represent values of M given by another strip, and the circles are values given by the ions from potassium sulphate. In all the measurements the strips were kept at the lowest temperature at which the currents were large enough for convenient measurement. This involved gradually increasing the temperature during any one experiment, on account of the decay of the emission with time.

It appears from these experiments that most of the ions given off by platinum have an electric atomic weight which is indistinguishable from that for the ions from potassium sulphate, that in the later stages ions for which M is very near to the sodium value (23·05) also appear, whilst finally, there are fleeting indications of the presence of ions with M in the neighbourhood of the atomic weight of iron (56).

The ions with M close to 40 were found not only to constitute the whole of the emission from freshly heated cleaned platinum. They were found also to carry the emission from a platinum strip when revived by heating in air and in a Bunsen flame, the greater part of the initial emission from iron and manganin strips, and the whole of the emission from a manganin strip due to revival by mechanical straining. Some o the data collected in various experiments to test the points mentioned are exhibited in the following table. Incidentally they demonstrate the very considerable accuracy with which ϵ/m and M can be measured with the apparatus used, and the high degree of consistency of the different experiments :—

Material and Treatment.	Duration of Observations.	Number of Measurements.	Extreme Values of M.	Mean Value of M.
Platinum, clean but not specially cleaned . .	5 hours	16	39·75—40·2	40·0
Platinum, cleaned with acids, etc. . . .	520 minutes	33	39·1—41·6	40·1
Platinum, cleaned with acids, etc. . . .	36 hours	—	38—40·1	39·1
Potassium sulphate .	280 minutes	10	39·2—41·1	40·2
Manganin strip, cleaned with reagents . .	4 hours	13	39·3—41·4	40·0
Iron, cleaned with reagents. . . .	55 minutes	4	39·8—40·0	39·9
Iron, cleaned with reagents. . . .	315 minutes	11	38·3—42·1	40·1
Platinum, emission restored by heating in air	160 minutes	11	37·9—39·8	39·0
Platinum, emission restored by heating in Bunsen flame . .	133 minutes	11	39·0—40·7	40·0
Manganin, emission restored by straining .	—	—	—	39·4

The table does not include the exceptional values already referred to, which were given by platinum strips which had been heated for a long time.

In the case of iron low values in the neighbourhood of the atomic weight of sodium were sometimes got at first, as well as values considerably above 40 in the last stages before the strips broke. These also have been omitted.

The experiments just described prove that the initial positive emission from hot bodies cannot be ascribed to charged atoms or molecules of any of the commoner gases which are likely to be present as impurities. The values of M for these are respectively :—

$$CO_+ = 28, \; H_+ = 1, \; H_{2+} = 2, \; CO_{2+} = 44, \; N_+ = 14, \; N_{2+} = 28,$$
$$O_+ = 16, \; \text{and} \; O_{2+} = 32.$$

The experimental determinations are much too accurate to admit of any of the bodies enumerated. The only admissible substances whose presence is at all likely are, $K_+ = 39.1$, $Ar_+ = 39.9$, or $Ca_+ = 40.07$. The mean values of M would agree better with Ar or Ca than with K ; but as the same is true for the ions from potassium sulphate, which there are good reasons for believing to be atoms of potassium, it seems most reasonable, on the evidence, to attribute these ions to the presence of potassium, and to assume that the method used tends to give values of M about 2 per cent too high.

NATURE OF THE IONS.

As the view which attributes the positive ions that carry the initial emission from hot bodies to the presence of gaseous contamination has acquired a good many adherents, it is perhaps desirable to consider a little more fully the arguments which have been advanced in support of such a position, and the reasons for considering them insufficient. The principal facts which have been held to support, or even by some authorities to establish, the gaseous origin of the ions in question, will be found enumerated in the following list :—

1. A wire which has lost the power of emission owing to continued heating in a vacuum is found to regain this property to a considerable extent if it is heated in a Bunsen flame or in air at atmospheric pressure.

2. A similar recovery takes place when the wire is exposed to various gases at a pressure of 50 to 100 atmospheres [1] at a relatively low temperature (about 200° C.).

---

[1] Z. Klemensiewicz, " Ann. der Physik," Vol. XXXVI, p. 796 (1911) ; cf. also p. 186, *ante.*

3. When most metals are first heated in a vacuum there is a considerable evolution of gas, the bulk of which usually consists of hydrogen, carbon monoxide, and nitrogen.

4. When a wire has been heated in a vacuum for a long time, so that the emission has become too small to measure, there is an emission in different gases, which seems to be a definite function of the nature and pressure of the gas (see Chapter VII). This phenomenon has been most completely studied in the case of platinum in an atmosphere of oxygen. The facts have been explained by the writer [1] on the hypothesis that the metal adsorbs or absorbs the gas, which it re-emits in the form of charged atoms. It has been suggested by various writers that the initial emission is an intensification of this phenomenon, owing to the presence of much larger amounts of gas in the original metal.

5. Horton [2] has found that carbon monoxide has a greater power of stimulating the emission of positive ions both from hot platinum and from heated salts than the other common gases, with the exception of hydrogen.

6. Determinations by Garrett [3] of e/m for the positive ions emitted by aluminium phosphate when heated, led him to conclude that about 10 per cent of them were charged atoms of hydrogen.

7. The experiments of Sir J. J. Thomson, described on p. 211, led him to conclude that the positive ions from hot platinum consisted of a mixture of charged atoms of platinum and of the surrounding gases. More recently Thomson [4] has returned to this question and has examined the positive ions from hot platinum by the same method as that used by him in investigating the positive rays. Most of the ions were found to have a value of M equal to 27, and he concludes that they are charged molecules of carbon monoxide. After the platinum had been soaked with hydrogen the average value of M was reduced to 9, indicating

[1] " Phil. Trans., A.," Vol. CCVII, p. 1 (1906).
[2] " Camb. Phil. Proc.," Vol. XVI, p. 89 (1911).
[3] " Phil. Mag.," Vol. XX, p. 582 (1910).
[4] " Camb. Phil. Proc.," Vol. XV, p. 64 (1908)

the presence of charged atoms or molecules of hydrogen also.

In weighing this evidence it is essential to realize that the portion enumerated under (1) to (5) is indirect so far as the *structure* of the ions is concerned. It may be of great importance in arriving at an understanding of the processes concerned in the emission of the ions, but it has only a secondary bearing on the question of material composition. For this purpose determinations of M give an immediate and final answer, provided that they are sufficiently accurate and do actually refer to the ions under discussion. At the same time it is necessary that the interpretation to which they lead should not be incompatible with the various points enumerated. I shall now show that none of the evidence really conflicts with the view that the *initial* positive ionization from hot metals consists of charged atoms of the alkali metals, and chiefly of atoms of potassium.

In the first place direct measurements of M for the emission from platinum revived by heating in air and in a Bunsen flame have consistently given values between 39 and 40, indicating that this emission has the same composition as that from a fresh wire, and that it does not consist of atoms or molecules of the various gases in which the heating has taken place. In conjunction with the fact that similar values are given by a strip revived by straining, this indicates that the effect of heating in gases consists in opening up the structure of the material and allowing access to the surface by alkaline impurities previously imprisoned. The effect of gases at high pressures is probably a direct mechanical one, although there are a number of actions which might affect the phenomena under these conditions. In any event there is no reason for presuming that the ions subsequently emitted consist of atoms or molecules of the gases used, in the absence of direct evidence to that effect and in the presence of direct evidence to the contrary.

With reference to (3) and (4) the writer [1] has measured simultaneously the quantity of gas and of positive electricity emitted on heating a new wire. Apart from the fact that both

---

[1] " Congrès de Radiologie," Liège, C.R., p. 50 (1905).

emissions were greatest at first and decayed with time, there was no evidence of a close correspondence between them. If the effect of heating in gases is due to the opening up of the metal by their solution or diffusion, (5) would be expected, as the gases carbon monoxide and hydrogen are notable for their power of diffusing into metals. Their chemical activity may also be a factor, as the positive emission from some salts has been found to be increased in the presence of reducing gases. The experiment of Garrett (6) has only an indirect bearing on the present question as it refers to a salt and not to a metal; but, in any event, it has not been confirmed as a fact by more recent and very careful experiments by Davisson.[1] The writer's experiments have afforded no evidence of the occurrence of hydrogen ions in the initial positive discharge from hot metals.

At first sight Thomson's experiments (7) appear to offer an immediate contradiction to the position now being maintained. His values of M are certainly quite different from those found by the writer, and there does not appear to be any likelihood that the differences can be attributed to errors in the measurements under comparison. There is, however, a very important difference in the conditions under which the experiments were made. Thomson used wires which had been heated for a long time in a vacuum before testing; so that presumably all, or almost all, the initial ionization would have been given off. Under these conditions there is no reason to expect that the values obtained would be those belonging to the carriers of the large initial emission.

If this is the explanation of the difference it follows that Thomson's measurements refer either to the positive emission due to the residual gases referred to under (4) or else to a permanent emission characteristic of the metal, and not, since the values of M are different, to remaining traces of the initial emission. Thomson's first experiments with platinum were made in an atmosphere of oxygen at 0·007 mm. pressure, and the majority of the ions were found to have a value of M near

[1] " Phil. Mag.," Vol. XXIII, p. 121 (1912).

14. We shall see in the next chapter that there is a considerable amount of indirect evidence indicating that the emission in an atmosphere of oxygen referred to under (4) is carried by charged oxygen atoms. This would agree well with the value obtained by Thomson. Again in the later experiments the value $M = 27$ was found at first, when the only gas whose presence could be detected spectroscopically was carbon monoxide, for which $M = 28$. After the platinum had been heated in hydrogen the average value of M was reduced to 9, indicating that some of the current was now carried by hydrogen ions. Thus these experiments furnish very definite evidence that when the initial emission has been got rid of, there is an emission in different gases of such a nature that the carriers are formed from the molecules of the gas.

In the first of the experiments referred to, Thomson also found indications of ions for which M had a value very close to the atomic weight of platinum. This rather indicates the existence in an old wire of an emission which is a property of the pure metal itself. Hitherto no other evidence of an emission of this kind has appeared, but it may be small and usually masked by the other kinds of emission to which reference has been made. The writer has several times attempted to detect the presence of ions for which M has values corresponding to the atomic weight of the metal or to the atomic or molecular weights of the traces of gas present, but has never succeeded in doing so. The method used was the same as in measuring M for the initial ionization. This method is admirable where currents of considerable size are available and where all the ions are alike. It is not suitable when the currents are small, as with wires which have been well glowed out, and it is quite incapable of detecting small quantities of one kind of ion mixed with large quantities of another. In these cases the methods used by Thomson have decided advantages. It would appear that there is room for more experiments on this subject, particularly by the canal ray method of measuring M.

In Chapter VIII we shall see that the view which attributes the large initial ionization to contamination by the alkali metals or their compounds receives indirect support from the very large emissions to which the compounds of these elements, and especially of potassium, give rise. This view also accounts for the large emission, described on p. 203, which is obtained when the walls of the glass tube containing the hot metal are heated. As there was no appreciable increase of gas pressure in this experiment the most likely agent would appear to be traces of salt vapours distilled from the glass.

THE QUANTITY OF ELECTRICITY EMITTED.

Since the emission from a fresh wire, heated at a constant temperature, after a while practically comes to an end there is a definite total quantity given off under these conditions. Nine wires,[1] each about 5 cm. long and 0·01 cm. diameter, heated to various temperatures between 600 and 800° C. all gave off about 10^{-5} coulomb. A strip[2] 0·1 cm. wide, about 1 cm. long, and whose weight was 0·055 gm. when heated at various temperatures up to 700° C., gave off about 2×10^{-6} coulomb. The amount obtained does not vary much with the temperature of heating, but there are some indications that it rises a little with rising temperature. The ratio of the mass of matter given off in the form of ions to the mass of the platinum heated is comparable with 10^{-7}, according to these numbers. The subject might repay further examination

[1] Liège, C.R., loc. cit.
[2] "Roy. Soc. Proc., A.," Vol. LXXXIX, p. 507 (1914).

CHAPTER VII.

THE EFFECT OF GASES ON THE LIBERATION OF POSITIVE IONS BY HOT METALS.

ALMOST all the experiments on the effect of gases on the emission of positive ions by metals have been made with platinum. This material has been used on account of its high melting-point, its chemical inertness, its mechanical suitability, and its other practical advantages, and not because it is believed to possess peculiar powers in respect to the phenomena under consideration. In fact it has been rather generally assumed that its behaviour may be regarded as typical of that of metals in general. As there is no certainty that this is the case it is probable that other metals would repay investigation in this direction.

The first systematic quantitative experiments bearing on the subject of this chapter were made by H. A. Wilson,[1] who measured the currents between two concentric cylindrical electrodes of platinum heated in air at atmospheric pressure. The inner electrode was a tube of 0·3 cm. outside diameter, and the outer electrode a tube of 0·75 cm. inside diameter. The heating was accomplished by placing the outer tube in a gas furnace. In this way the temperature of the electrodes could be kept within 5° of any desired temperature up to 1400° C. In general the inner electrode was the colder, but the difference of temperature could be reduced by blowing a current of air in the space between them. By blowing cold air down the inside of the inner tube its temperature could be made much lower than that of the outside electrode.

The currents were measured under varying potential differences at a constant temperature and under constant potentials

---

[1] "Phil. Trans., A.," Vol. CXCVII, p. 415 (1901).

at different temperatures. With a given potential difference and temperature it was found that the current was always greatest, except at very low potentials, when the outer tube was positive. With the outer tube positive there was no indication of an approach to saturation except when the inner tube was cooled by blowing air through it. With the outer tube negative the current exhibited very little increase between 100 and 400 volts except when the inner tube was artificially cooled. Under constant potential differences the currents increased rapidly with rising temperature, of which the logarithm of the current was very close to a linear function.

With the arrangement used there are a large number of possible sources of ionization. These include volume ionization of the hot air, formation of positive and negative ions by interaction between the gas and the electrodes, and the emission of ions of both signs from the electrodes which would take place if no air were present. Inasmuch as the currents with large potential differences were much greater when the larger and hotter electrode was positive it is reasonable to suppose that the observed effects were mainly due to the emission of positive ions by the hot platinum, either of itself or by interaction with the gas. We have seen that the emission from freshly heated platinum wires decays with time, rapidly at first and then more slowly, when the wires are heated in a vacuum. Wilson noticed effects analogous to both of these when the tubes were heated at atmospheric pressure. The readings taken before the quick decay had occurred were disregarded; but inasmuch as the currents at constant temperature and potential fell off slowly but continuously during the course of the experiments it seems unlikely that the effects due to the initial emission from the hot metal were eliminated. This conclusion is supported by the large magnitude of the current-densities obtained. These were easily measured with a galvanometer, and were very much larger than those obtained later by the writer when a well-glowed-out platinum wire was heated in air at atmospheric pressure (see p. 235).

By assuming that the ions were formed by the dissociation

15 *

of the gas at the surface of the platinum and applying the formula [1]

$$\frac{q}{2}\left(\frac{1}{T_2} - \frac{1}{T_1}\right) = \log \left\{\frac{x_1^2}{x_2^2}\frac{1 - x_2^2}{1 - x_1^2}\frac{T_{21}}{T}\right\},$$

where q is the heat of dissociation, and x_1 and x_2 are respectively the fractions of the gas dissociated at the absolute temperatures T_1 and T_2, Wilson found values for q in the neighbourhood of 60,000 calories. This investigation is noteworthy as it forms the first attempt to estimate the energy changes involved in the formation of the ions, from the rate of variation of the currents with temperature. More recently Wilson [2] has shown that his numbers for the approximately saturated currents obey the formula $i = A T^{\frac{1}{2}} e^{-b/T}$ with $b = 25,000$ degrees centigrade.

The effect of oxygen, air, nitrogen, helium, and hydrogen on the emission of positive ions from platinum has been examined in some detail by the writer. [3] A thin platinum wire, electrically heated and arranged in the form of a loop, was mounted in a glass tube alongside an insulated platinum plate, and the current which flowed from the wire to the plate was measured. Considerable attention was paid to the cleanliness of the tube and the purity of the platinum and of the gases used. Before commencing the measurements the platinum wires were glowed out for long periods in an oxygen vacuum until the initial positive emission had become very small and showed no appreciable recovery on standing. On letting in small quantities of oxygen and other gases it was then found that there was a small emission which was a definite function of the pressure and nature of the gas and of the temperature of the wire. The magnitude of the currents dealt with is indicated by the following numbers which refer to one of the earlier experiments made before the initial emission had entirely disappeared. The wire under test was 7 cm. long and 0·01 cm. in diameter. The positive emission on first

[1] Van't Hoff, "Lectures on Theoretical and Physical Chemistry," Vol. I, p. 145.
[2] "Phil. Trans., A.," Vol. CCVIII, p. 248 (1908).
[3] *Ibid.*, Vol. CCVII, p. 1 (1906).

heating at 804° C. was found to be 1.62×10^{-8} amp., the pressure given by the McLeod gauge being 0·00005 mm. This current decayed to one-half its value in 10 minutes and to one-tenth in about an hour. Even after heating *in vacuo* for several hours a day, at temperatures in the neighbourhood of 800° C., for about two weeks the wire still gave small currents under the best available vacuum conditions. Thus at 0·0003 mm. pressure a saturation current of 9.6×10^{-13} amp. was obtained at 721° C. when the wire was charged positively. On letting in oxygen to a pressure of 0·045 mm. and keeping the temperature constant the current increased to 1.8×10^{-12} amp. It was found that the small current which did not depend on the pressure of the gas gradually disappeared with continued heating, whereas the additional current caused by the gas did not. It thus appears that with wires which have been heated in a vacuum for a long time there is a positive emission which is a definite function of the pressure of the surrounding gas.

Under the conditions of these experiments saturation was attained, except in some of the gases at high pressures, by the application of moderate potentials. Even in the exceptional cases there was a close approach to saturation. Most of the experiments were made at temperatures so low that there was no current when the wire was charged negatively. In thèse experiments, therefore, the currents are due entirely to ions emitted by the hot electrode. None of the observed effects are due to volume ionization of the gas or to the emission of ions of opposite sign from the collecting electrode. The conditions are thus much simpler than where two hot electrodes are employed with hot gas in between, and the results are correspondingly easier of interpretation. Inasmuch as the initial emission was allowed to decay before the measurements were made, and the currents measured only appeared when gas was admitted and disappeared when it was removed, it is clear that these effects are something quite different from the initial emission from freshly heated wires which was considered in the last chapter. We shall now proceed to describe in more detail the phenomena exhibited in the different gases.

1. *Oxygen.*—The currents with the wire charged positively were found to saturate very readily at all pressures up to atmospheric. At high pressures and low temperatures the emission exhibited a curious instability; the current under apparently constant conditions kept suddenly increasing to a temporary high value and then returning to about the original level. The cause of this instability has not been discovered, but on the assumption that it is due to some secondary pheno- menon, the minimum values of the currents were taken to represent those due to the direct action of the gas. This diffi- culty was not encountered to any considerable extent at low

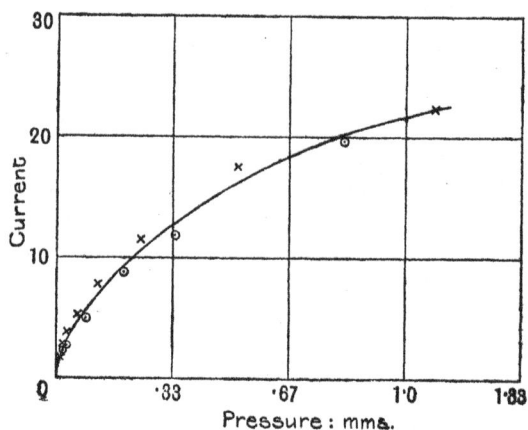

Fɪɢ. 25.

pressures at any temperature or at high pressures when the temperature was high.

At low temperatures the saturation current was nearly pro- portional to the square root of the pressure, when this was under 1 mm. At higher temperatures, in the neighbourhood of 1100° C. to 1200° C., the current was almost proportional to the pressure over this range. At all the temperatures there was very little variation of current with pressure at high pres- sures (200 to 800 mm.). The behaviour at 828° C. for pres- sures below 1 mm. is shown in Fig. 25.

The variation with temperature of the saturation current at a constant pressure of 1·47 mm. was also examined. The superficial area of the hot wire used was 0·223 sq. cm. and its diameter 0˙01 cm. The currents obtained are shown in the next table. The electron currents with the wire charged to − 40 volts are added for comparison. These are probably somewhat greater than the corresponding saturation currents owing to the occurrence of some ionization by collision.

Temperature (Centigrade).	Positive Emission (Amperes).	Electron Current (Amperes).
708	$1˙6 \times 10^{-12}$	—
770	$6˙7 \times 10^{-12}$	—
826	$1˙5 \times 10^{-11}$	—
883	$3˙2 \times 10^{-11}$	$1˙1 \times 10^{-14}$
940	$5˙8 \times 10^{-11}$	$6˙7 \times 10^{-14}$
999	$1˙1 \times 10^{-10}$	$8˙0 \times 10^{-13}$
1058	$3˙8 \times 10^{-10}$	$6˙2 \times 10^{-12}$
1119	$6˙4 \times 10^{-10}$	$3˙2 \times 10^{-11}$
1181	$1˙1 \times 10^{-9}$	$3˙3 \times 10^{-10}$
1227	$1˙7 \times 10^{-9}$	$1˙6 \times 10^{-9}$

At the lower temperatures the negative emission is negligible compared with the positive but increases more rapidly with the temperature ; so that at this pressure the two currents are equal at about 1230° C. The values for both emissions satisfy the equation $i = AT^{\frac{1}{2}}e^{-b/T}$ with different values of the constants. The value of b for the data referring to the positive emission is $1˙52 \times 10^{4}$ ° C. The values of the positive emission per unit area from four wires of various sizes which had undergone different treatment were determined at 1˙5 mm. pressure at 770° C. and 880° C. They were found all to be the same at the same temperature within the limits of the experimental error, indicating that the emission caused by oxygen is due to the platinum itself and not to some adventitious impurity. One of the wires was subsequently heated strongly in hydrogen and the treatment was found to reduce the permanent emission in oxygen by a factor of nearly four. This effect may, however, be caused by a change in the crystalline structure of the platinum which is known to be brought about by this treatment.

When the temperature of the wire was kept constant and the pressure of the oxygen raised it was found that the emission was too low at first and gradually increased to the final steady value. Similarly, when the pressure was reduced the observed emission was too high initially. These phenomena indicate that the emission is not due directly to an interaction between the hot metal and the surrounding gas, but rather that it arises from the gas which is either dissolved in, combined with, or adsorbed by the metal. Any of these processes would be expected to take some time to reach a condition of equilibrium after an alteration in the pressure of the gas had been made. The definitely established facts as to variation of this emission with temperature and pressure can be accounted for if we adopt the hypothesis that the emission at constant temperature is proportional to the number of oxygen *atoms* held in the surface layer of the platinum at any instant. This hypothesis may be formulated quantitively as follows :—

Let a be the total number of platinum atoms per unit area of the surface which are available for combination with oxygen, and let x be the number which are combined with oxygen atoms at any instant ; the number of free platinum atoms is then $a - x$. If p is the partial pressure of the dissociated oxygen, the free oxygen atoms will become entangled at a rate proportional to $p \times (a - x)$, and will be liberated from the surface layer at a rate proportional to x. Thus, if t is the time,

$$\frac{\partial x}{\partial t} = Ap(a - x) - Bx,$$

where A and B are constant at a given temperature. The state of equilibrium is determined by $\frac{\partial x}{\partial t} = 0$; so that

$$x = \frac{Apa}{Ap + B} = \frac{af(P)}{b + f(P)} \qquad . \qquad . \qquad . \quad (1)$$

where $b = B/A$ and $p = f(P)$, P being the total pressure of the external oxygen. On this theory the emission is proportional to x multiplied by a factor of the form $AT^{\lambda}e^{-b_1/T}$, with A, λ, and b_1 constant, representing the rate of the reaction by which the positive ions are ejected. At high pressures $f(P)$ becomes large compared with b ; so that x approaches a asymptotically at all temperatures. The emission will thus become independent of the pressure at high pressures whatever the temperature. This was found to be the case. At low temperatures the dissociation will be small even at low pressures ; so that $f(P)$ will be nearly proportional to $P^{1/2}$. Thus at low temperatures and pressures the emission should be closely represented by the formula

$$i = \frac{a\mathrm{P}^{1/2}}{\beta + \mathrm{P}^{1/2}} \qquad . \qquad . \qquad . \qquad . \qquad (2)$$

where a and β are constant at constant temperature. As a matter of fact, if this formula holds for low pressures, it would be expected to be fairly near the truth at high pressures also, since the precise form of $f(\mathrm{P})$ makes very little difference when P is large. That this formula covers the experimental values almost within the limits of experimental error, at low temperatures, for the range from 0·003 mm. to 760 mm., is shown by the numbers in the next table. The observations were made at 820° C.; the calculated numbers were deduced from (2), using $a = 5\cdot6 \times 10^{-11}$ amp., and $\beta = 4\cdot0$ (mm. of mercury)$^{1/2}$.

Pressure P. Mm. of Mercury.	P$^{1/2}$.	Calculated Emission $i = a\mathrm{P}^{1/2}/(\beta + \mathrm{P}^{1/2})$.	Observed Emission.
0·003	0·055	0·75	1·0
0·017	0·41	5·2	5·9
1·5	1·22	13·2	15
3·1	1·76	17·1	17
6·1	2·47	21·3	20·7
10·7	3·27	25	23·5
17·0	4·12	28·4	26·5
30	5·48	32·3	30
53	7·28	36·5	34
97	9·85	39·5	38
200	14·3	43·7	43
399	20	46·7	49·3
587	24·2	48·3	50·5
766	27·7	49	53·5

Experiments at a lower temperature (730° C.) gave at least as good an agreement with (2), using $a = 1\cdot2 \times 10^{-11}$ amp. and $\beta = 3\cdot9$ (mm. of mercury)$^{1/2}$.

At a higher temperature (1170° C.) the experimental values did not agree with equation (2) at low pressures, but were found to be in excellent agreement with

$$i = \gamma \mathrm{P}/(\delta + \mathrm{P}) \qquad . \qquad . \qquad . \qquad (3)$$

from 0·14 to 89 mm., with $\gamma = 3\cdot8 \times 10^{-9}$ amp. and $\delta = 4\cdot8$ mm. of mercury. This result is to be expected if the greater part of the oxygen near the surface of the metal is dissociated at this temperature and at the lower pressures, because $f(\mathrm{P})$ would then be more nearly proportional to P than to P$^{1/2}$. It is only at the lowest pressures that the dissociation need be relatively complete as the formula is not very sensitive to the form of $f(P)$ at the higher pressures.

Nitrogen.—At low temperatures the positive emission in this gas was smaller than in oxygen and more nearly comparable with the negative emission. Perhaps for this reason higher voltages were required to attain saturation than in oxygen at similar pressures when the wires were positively charged. This is shown by the upper curve in Fig. 26 which represents the relation between current and electromotive force for the positive emission in nitrogen at atmospheric pressure and 920° C. When the pressure was changed at constant temperature the emission increased rapidly with

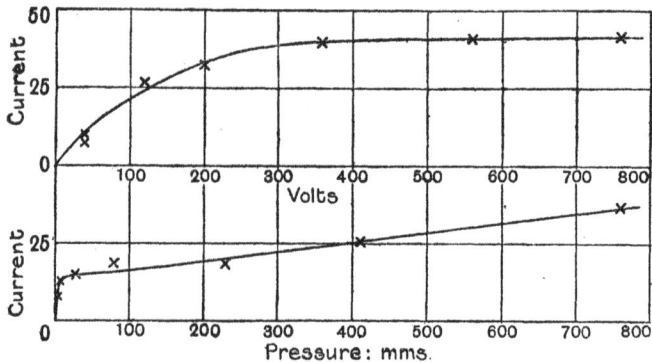

FIG. 26.

rising pressure at low pressures. The rate of increase of emission with pressure diminished very greatly at higher pressures but not so much so as with oxygen. There was no tendency for the emission to approach a constant value at pressures in the neighbourhood of atmospheric, but a regular linear increase of emission with pressure was observed from about 30 to 800 mm. The variation of current with pressure at 920° C. is shown in the lower curve of Fig. 26. In these experiments there was a small current at low pressures which was independent of the pressure.

The variation of the saturation current with temperature in nitrogen at 2·8 mm. pressure is shown by the following numbers :—

Temperature (Centigrade).	Saturation Currents (Amperes per sq. cm.).	
	+ *ve* Emission.	- *ve* Emission.
827	$3 \cdot 0 \times 10^{-13}$	$4 \cdot 4 \times 10^{-14}$
900	$1 \cdot 7 \times 10^{-12}$	$5 \cdot 8 \times 10^{-13}$
907	$3 \cdot 8 \times 10^{-12}$	$1 \cdot 5 \times 10^{-12}$
984	$2 \cdot 76 \times 10^{-11}$	$3 \cdot 5 \times 10^{-11}$
1071	$9 \cdot 9 \times 10^{-11}$	$4 \cdot 7 \times 10^{-10}$

Both the positive and negative emissions follow the formula $i = AT^{\frac{1}{2}}e^{-b/T}$; for the positive emission the value of b is $3 \cdot 56 \times 10^{4 \circ}$ C. and for the negative $5 \cdot 6 \times 10^{4 \circ}$ C. The positive emission increases less rapidly with temperature than the negative emission, but more rapidly than the positive emission in oxygen. Thus, if the same laws continue to operate, the positive emission in nitrogen should exceed that in oxygen at high temperatures. The temperature at which the two emissions become equal should be lower at atmospheric pressure than at pressures of about 1 mm.

Air.—With some reservations the behaviour of the positive emission in air is intermediate between that in oxygen and that in nitrogen. Thus when the electromotive force is varied the approach to saturation is observed at a lower potential than in nitrogen but at a higher potential than in oxygen. Again the saturation current at constant temperature increases with pressure at high pressures less rapidly than in nitrogen but more rapidly than in oxygen. The values of the currents at atmospheric pressure from a wire whose effective area was 0·66 sq. cm. are shown in the next table. The currents were approximately saturated.

Temperature (Centigrade).	Current (Amperes).	
	Positive Emission.	Negative Emission.
812	$9 \cdot 3 \times 10^{-12}$	—
893	$2 \cdot 2 \times 10^{-11}$	$3 \cdot 3 \times 10^{-14}$
900	$5 \cdot 2 \times 10^{-11}$	$5 \cdot 3 \times 10^{-14}$
978	$3 \cdot 3 \times 10^{-10}$	$3 \cdot 2 \times 10^{-13}$
1064	$8 \cdot 0 \times 10^{-10}$	$4 \cdot 2 \times 10^{-12}$
1150	$2 \cdot 0 \times 10^{-9}$	$3 \cdot 8 \times 10^{-11}$
1236	$6 \cdot 7 \times 10^{-9}$	$2 \cdot 6 \times 10^{-10}$

These currents follow the usual formula for the temperature variation; the values of b being, for the positive emission, $2 \cdot 46 \times 10^{4 \circ}$ C., and for the negative, $4 \cdot 49 \times 10^{4 \circ}$ C. The value for the positive emission in oxygen at atmospheric pressure was $1 \cdot 52 \times 10^{4 \circ}$ C., and for nitrogen, $3 \cdot 56 \times 10^{4 \circ}$ C. Thus the value of b for the positive emission in air at atmospheric pressure is nearly equal to the mean of the values for oxygen and nitrogen at the same pressure.

Although the foregoing considerations show that the phenomena observed in air are to a certain extent a compromise between those exhibited in nitrogen and oxygen respectively, the emission in air is not equal to the sum of the effects due to the nitrogen and oxygen present supposed to act independently. Thus at a certain temperature the wire, when heated in the gases at atmospheric pressure, gave emissions proportional to the following numbers: nitrogen 29, oxygen 290, air 70. The emission in oxygen at a pressure equal to its partial pressure in the atmosphere is only about 10 per cent less than at atmospheric pressure; so that the oxygen alone which was present in the air should have given an emission of about 260 in the units used. Adding in the effect of the nitrogen, the principle of simple superposition of the effects would give an emission equal to about 280, instead of the observed value 70. It thus appears that the nitrogen does not act merely as a diluent of the oxygen, but has a distinct inhibiting effect on the more active gas. It is possible that it does this by combining with platinum atoms which would otherwise be free to take up oxygen; but there are clearly a number of other ways in which the observed effects might arise.

Helium.—Only a few experiments have been made with this gas; but the results are of particular interest since they indicate the existence of a positive emission caused by a gas which is believed to be incapable of acting chemically on the hot body. The gas was freed from impurities by subjecting it, in a tube connected with the testing bulb, to a luminous discharge from a cathode of sodium potassium alloy. After this treatment the following values of the currents at different pressures were observed when the emission had assumed a steady condition at 907° C. :—

Pressure (mm. of mercury) . . .	0·07	0·32	2·4
Current ($1 = 3·3 \times 10^{-14}$ amp. per sq. cm.)	18	54	130

At this temperature and at a pressure of 2 mm. of mercury the positive emission in helium appears to be about twice that in nitrogen and one-fortieth that in oxygen under like conditions.

Hydrogen.—The emission in this gas at low temperatures is difficult to investigate, as it changes very slowly with time after any alteration in the conditions has been made, and the final equilibrium value takes an inordinate length of time to become established. At 900° C. and 3·8 mm. pressure the steady positive emission was found to be 7×10^{-12} amp. cm.$^{-2}$. This is about one twenty-fourth of the value in oxygen at the same temperature and pressure. At 1300° C. H. A. Wilson[1] found the following values for the currents from a positively charged wire at different pressures :—

Pressure (mm. of mercury)	9	156	766
Current ($i = 10^{-9}$ amp.)	4	24	40

Thus at this temperature the positive emission from platinum in hydrogen appears to resemble that in oxygen and nitrogen in so far as it varies only rather slowly with the pressure at high pressures.

The following values of the approximately saturated currents at a pressure of 1·9 mm. and different temperatures were found by the writer :—

Temperature, centigrade	860	1017	1181
Positive emission (amps. per cm.²)	$2·5 \times 10^{-11}$	$1·3 \times 10^{-10}$	$1·12 \times 10^{-9}$
Negative „ „ „ „	$4·7 \times 10^{-10}$	—	$1·1 \times 10^{-5}$

The values of b calculated from these data are $1·79 \times 10^{4}$° C. for the positive emission and $4·74 \times 10^{4}$° C. for the negative. Experiments at a higher pressure (226 mms.) led to the following numbers :—

Temperature, centigrade.	860	1017	1097	1181
Positive emission (amps. per cm.²)	$4·1 \times 10^{-11}$	$3·8 \times 10^{-10}$	—	$1·4 \times 10^{-8}$
Negative emission (amps. per cm.²)	$1·0 \times 10^{-7}$	—	$1·25 \times 10^{-5}$	$2·8 \times 10^{-5}$

From these data $b = 2·85 \times 10^{4}$° C. for the positive emission and $2·78 \times 10^{4}$° C. for the negative. In hydrogen the value of b for the positive emission increases with increasing pressure, thus exhibiting the contrary behaviour to that for the negative emission.

[1] " Phil. Trans., A.," Vol. CCII, p. 243 (1903).

Hydrogen Diffusing into Air.

The positive emission from platinum in air when hydrogen is diffusing out of the platinum has been examined by the writer.[1] The apparatus consisted of an electrically heated platinum tube with a coaxial cylindrical electrode. The rate at which the hydrogen diffused out of the platinum was controlled by varying the pressure of the hydrogen inside the tube. Under these conditions the quantity of hydrogen which diffuses out has been shown[2] to be proportional to the square root of the pressure, at constant temperature. Most of the experiments were made at about 1200° C. At low temperatures (800° C.) the currents when the wire was positively charged showed no sign of approaching saturation up to 960 volts. This indicates that the platinum had been insufficiently glowed out and was behaving like a fresh wire (see p. 200). At the higher temperatures (about 1200° C.) approximate saturation was attained with 80 volts, but even here changes of the current with time were observed after altering the potential. Such changes have generally been found to be characteristic of the behaviour of freshly heated wires. It may be important to remember that the platinum was in this condition as it is possible that a well-glowed-out tube might behave differently.

The variation, with the pressure of the hydrogen inside the tube, of the positive emission at 1200° C. is shown by the following numbers :—

Pressure P (mm.).	Positive Emission ($i = 1.8 \times 10^{-11}$ amp. per cm.^{2}). Found.	Calculated.
0	42	42
30	51	52
60	56·3	55
172	64	65·6
780	90	92·4

The numbers in the last column were calculated by assuming that the emission was equal to $a + bP^{\frac{1}{2}}$, a and b being constants. The agreement of the results shows that the emission consists of two parts, one proportional to the square

[1] Loc. cit., p. 57.
[2] Cf. Richardson, Nicol, and Parnell, " Phil. Mag.," Vol. VIII, p. 1 (1904).

root of the pressure of the hydrogen inside the tube and the other independent of it. The effect of the hydrogen diffusing out of the platinum therefore is to produce an additional number of positive ions proportional to the quantity of hydrogen which diffuses out. Other considerations make it probable that the hydrogen inside the platinum is in the atomic state, and these results indicate that part, at least, of the dissolved atoms are ionized. The values of the currents show that only a small fraction (about 10^{-7}) of the escaping hydrogen gets away in the form of ions ; but the proportion in the interior which is in the ionic state may be much higher.

The positive emission from the tube at different temperatures was measured, both when the interior of the tube was evacuated and when it was filled with pure hydrogen at atmospheric pressure. The difference gives the emission due to the diffusing hydrogen at the different temperatures. The values obtained for this quantity are shown in the following table :—

Temperature, centigrade	973	1052	1129	1200	1262	1331
Emission due to hydrogen ($i = 1.8 \times 10^{-12}$ amp. per cm.^{2})	6	17	43	80	172	340

These numbers increase more rapidly with the temperature than the quantities of hydrogen diffusing out. Hence the efficiency for liberating positive ions of a given amount of hydrogen diffusing out of platinum increases with rising temperature.

EFFECTS CAUSED BY CHANGING FROM ONE GAS TO ANOTHER.

When a wire has been heated for a long time in one gas the positive emission on admitting small quantities of a second gas is larger, at first, than the steady value to which it ultimately settles down.[1] This effect appears to occur even if the wire is heated for a long time in a good vacuum with the pump in continuous operation before the second test is made. Under these conditions the initial excess of the emission over the final value is larger than the residual emission before the new gas is admitted. These effects have been observed in oxygen,

---

[1] Richardson, "Phil. Trans., A.," Vol. CCVII, p. 1 (1906).

nitrogen, helium, and hydrogen. The decay of the positive emission in hydrogen was accompanied by a parallel increase in the negative emission. This effect has not been looked for with the other gases. In hydrogen the flow of the negative emission apparently caused a small temporary increase in the positive emission on subsequent testing. The decay of the positive emission after admitting a new gas, and its temporary revival by allowing the negative emission to flow, are similar to the effects observed when fresh wires are heated; but the currents in these experiments are enormously smaller than those given by fresh wires at the same temperature.

NATURE OF THE IONS.

So far no direct evidence as to the nature of the ions which carry the positive emission from hot metals caused by gases has been adduced; although all the results described appear to harmonize with the view that the carriers are charged atoms or molecules of the surrounding gas liberated, perhaps somewhat indirectly, by its interaction with the metal. In discussing the experiments in oxygen we assumed quite explicitly that the ions were charged oxygen atoms liberated by the gas present in the surface layers of the platinum. This assumption was seen to give a fairly straightforward quantitative explanation of the phenomena. It is questionable whether they could be accounted for by any other equally simple hypothesis.

As we have already seen, the measurement of the electric atomic weight M of the ions is the only completely satisfactory way of settling their nature. The measurement of M for these ions presents some difficulty on account of the smallness of the currents when the pressure of the gas causing them is not too large seriously to interfere with the motion of the ions. The only measurements of e/m or M which can have any claim to apply to the positive emission caused by a gaseous atmosphere are those of Thomson, referred to on p. 221, since they are the only ones which have been made with wires sufficiently well-glowed-out to get rid of the initial emission. With platinum in an air vacuum Thomson found, using the cycloid method, ions for which M was near 14. This agrees with the view that the ions are atoms of oxygen or nitrogen or both. In the later experiments the more accurate positive-ray method of measuring M was used. With platinum in a carbon monoxide vacuum ions for which M = 27 were found. For molecules of CO with a single charge M would be 28. After saturating the platinum with hydrogen the average value of M found subsequently fell to 9. These results clearly afford strong support to the view that the ions which constitute this emission at low pressures are charged atoms or molecules of the gases concerned. At the same time only a limited number of experiments have been made under the conditions utilized by Thomson, and the reasons here adduced for regarding the emissions as of the same character as those described in the present chapter are perhaps not altogether conclusive; so that there appears to be room for further investigation in this direction.

There is definite evidence that the ions which carry the positive emission in air at atmospheric pressure are of atomic magnitude, when the platinum has not been glowed out in a vacuum so as entirely to get rid of the initial emission. H. A. Wilson [1] has shown that the main features of the current-E.M.F. curves obtained by him, and referred to at the beginning of this chapter, can be accounted for on the hypothesis, that the only conditions which prevent the attainment of saturation are the mutual repulsion of the ions, which alters the distribution of the field between the electrodes, and the backward diffusion of the ions into the emitting electrode. When the potential difference is neither too high nor too low, the disturbance of the electric field is small but not negligible. In this part of the curve the potential difference V satisfies the equation

$$V = \frac{i}{\zeta_2 k} \cdot \frac{\log r_2 - \log r_1}{2\pi} + \pi \left(2r_2^2 \log \frac{r_2}{r_1} - r_2^2 + r_1^2 \right) \zeta, . \quad (4)$$

where i is the current, k the mobility of the ions, ζ_2 the volume density at the electrode whose radius is r_2, and r_1 the radius at the other electrode. Thus, this part of the i, V curve is a

[1] " Electrical Properties of Flames and Incandescent Solids," p. 38.

straight line whose intercept on the voltage axis is proportional to ζ_2. By measuring the intercept and the slope of the line, together with a knowledge of the radii of the cylinders, both ζ_2 and k may be determined. The value of k for the positive ions found by this method in air at atmospheric pressure at 1000° C. is 43 cm. sec⁻¹ per volt cm.⁻¹. This agrees with the value to be expected for a positively charged atom of potassium from kinetic theory considerations, so far as the data for the calculation of this quantity can be relied on. Whether this interpretation is accepted or not, the experimental value of k leaves little room for doubting that the ions are charged atoms or molecules.

This conclusion only applies, generally speaking, to the ions as they are found close to the hot metal. In the colder gas at some distance away the ions grow in size, and the value of k diminishes. This is shown by the experiments of McClelland described on p. 6, in which the mobility of the ions in the gas drawn away from incandescent metals was measured directly by a blowing method. It is also shown by some experiments made by Rutherford,[1] who measured the current from a hot platinum plate to a parallel electrode when the currents were very small, compared with the saturation value. Under these conditions the relation between the voltage V and the current density i is expressed by the equation

$$V = \frac{2}{3}\left(\frac{8\pi i}{k}\right)^{\frac{1}{2}} l^{\frac{3}{2}} \qquad . \qquad . \qquad . \qquad (5)$$

where k is the mobility of the ions, and l the perpendicular distance between the electrodes. It is clear that k may be deduced by measuring V, i, and l. Rutherford found that k increased with increasing distance between the plates. In addition, very small values of k were found at high temperatures. These were attributed to the loading of the ions by the platinum dust which is sputtered under these conditions.

We have seen that the emission of positive ions from platinum in various gases satisfies the equation $i = AT^{\frac{1}{2}}e^{-b/T}$. The values of the constants in a number of cases are collected

---

[1] "Phys. Rev.," Vol. XIII, p. 321 (1901).

in the following table.   The values for the negative emission under similar conditions are added for comparison :—

Gas.	Pressure (mm.).	A_+	b_+	A_-	b_-
Oxygen . .	2	7×10^{15}	$1 \cdot 52 \times 10^4$	4×10^{28}	$6 \cdot 78 \times 10^4$
Air . . .	760	7×10^{18}	$2 \cdot 46 \times 10^4$	10^{21}	$4 \cdot 49 \times 10^4$
Nitrogen .	2·8	4×10^{21}	$3 \cdot 56 \times 10^4$	3×10^{26}	$5 \cdot 6 \times 10^4$
Hydrogen .	1·9	10^{16}	$1 \cdot 79 \times 10^4$	10^{23}	$4 \cdot 74 \times 10^4$
Hydrogen .	226	10^{20}	$2 \cdot 85 \times 10^4$	3×10^{19}	$2 \cdot 78 \times 10^4$

It is a remarkable fact that the constants for the positive emission, in the table above, exhibit a linear relation between log A and b similar to that shown by the constants for the negative emissions from platinum and tungsten, which was considered in Chapter IV.   Moreover, the constant a/c, or c', considered on p. 137, has a very similar value, being equal to $1 \cdot 43 \times 10^3$ for the positive emission from platinum in various gases, as compared with the values $3 \cdot 29 \times 10^3$ for the negative emission from platinum and $2 \cdot 56 \times 10^3$ for the negative emission from tungsten.   We have seen that, in the case of the negative emissions, the linear relation in question is closely connected with the contact difference of potential between the metal contaminated by gases and the pure metal.   A relation between the constants A and b for the positive emission, such as is contained in the data above, would be expected to arise, in the same way as for the negative, if the positive ions were present in the metal and if their internal concentration were independent of the nature and pressure of the gas and of other external factors.   The theory of the emission of these positive ions would then be similar to that of the emission of the negative electrons.   The main differences would result from the atomic dimensions of the positive ions and their much smaller concentration in the metal.   The effect of gases on the positive emission would then be closely connected with the corresponding contact potentials, although the effects might not show an exact correspondence with those given by the negative emission on account of the atomic character of the positive ions.   With the positive ions there may be a material, as well as an electrical, factor to consider.   Several years ago the writer [1] pointed out that the phenomena which characterize

[1] " Phil. Trans., A.," Vol. CCVII, p. 61 (1906).

the positive emission from " old " platinum wires in various gases could be united into a coherent whole from this point of view.   At that time, however, such a theory was considered improbable from the fact that gases like oxygen appeared to exert an effect on the positive emission out of all proportion to that exerted on the negative.   It may be, however, that the issue is not so simple as was supposed, and that the hypothesis under consideration has been dismissed too lightly.   On the other hand, if the hypothesis is accepted some other explanation will have to be sought for the values of e/m found by Thomson which, as we have seen, make the emitted ions atoms of the surrounding gas.   Moreover, the linear relation between the constants A and b for the positive ions rests only on five pairs of values, and the agreement may prove to be accidental. It is clear that this subject is one which affords scope for further experimental investigation.

THE EMISSION FROM FRESH WIRES IN GASES.

When platinum is freshly heated in air at pressures up to atmospheric the emission, particularly at rather low temperatures, exhibits interesting peculiarities, which show a close resemblance to some of the effects observed with freshly heated wires in a vacuum.   The phenomena to be described refer to platinum heated in air at atmospheric pressure unless the conditions are definitely stated to be otherwise. H. A. Wilson [1] observed that the positive emission decayed with the time of heating, rapidly at first and then more slowly.   The writer [2] found that at moderate temperatures this decay increased rapidly with the positive potential applied to the hot metal and was inappreciable when the latter was earthed, or at a relatively low potential.   Thus a new wire at 925° C. was found when charged with + 40 volts to give a current of 100 divisions which remained constant for 100 minutes.   On raising the potential to + 760 volts the current had the following values at the times stated :—

[1] " Phil. Trans., A.," Vol. CXCVII, p. 415 (1901).
[2] *Ibid.*, Vol. CCVII, p. 30 (1906).

Time minutes .	.	0	3	6	9	14	20	25	36	47	54	60	66
Current (divisions)	.	3570	1930	950	760	570	485	475	190	115	112	103	103

On returning to + 40 volts the currents at successive intervals of 6 minutes were 80, 84, 90, and 94 divisions. This experiment was made with a thin wire of about 0·01 cm. diameter surrounded by a coaxial cylindrical electrode of 3·2 cm. diameter provided with guard rings. Similar results were obtained when the thin wire was replaced by a heated platinum tube of 0·2 cm. outside diameter. It was also noticed that the positive emission *increased* in magnitude if the hot electrode was left negatively charged.

The diminution of the rate of decay of the emission caused by reducing the applied positive potential has been confirmed by the observations of W. Wilson [1] and of Sheard.[2] The latter also observed that the emission from a positively charged wire at a low temperature could be increased by heating the wire to a higher temperature for some time in a negatively charged or uncharged condition. At 628° C. he found that a wire under test gave an emission of 14 divisions under + 200 volts which showed no appreciable decay with time. The wire was then connected to earth and heated during intervals of 10 minutes at various temperatures up to 840° C. Subsequent to each of these heatings the emission under + 200 volts was tested at the original temperature of 628° C. It was found to be greatly increased by the treatment. The increased emission was a definite function of the temperature at which the wire had been heated under zero voltage, with sharp maxima at 650° C. and 760° C. respectively and a minimum between. The current at 628° C., after heating to 760° C., was about 40 times as great as that observed prior to this treatment. Similar, but smaller, effects with maxima at the same points were observed with a wire which had been revived by heating in a Bunsen flame.

The fact that the emission decays most rapidly when a large positive potential is applied to the hot metal shows that the removal of charged ionizable matter by the electric field

---

[1] " Phil. Mag.," Vol. XXI, p. 634 (1911).
[2] *Ibid.*, Vol. XXVIII, p. 170 (1914).

is an essential feature of the decay phenomenon. If this material is not removed by the field it diffuses back to the hot metal and helps to emit more positive ions. It seems fairly clear that part, at any rate, of the active material is not available at relatively low temperatures but is only formed at somewhat higher temperatures ; so that the effect of heating alone may be, in certain cases, to increase and not to diminish the current at a standard temperature. The point has not been investigated very carefully, but the writer's impression is that the current decays rapidly under heating alone at very high temperatures. It is probable that under these conditions the heating destroys the active substance formed at intermediate temperatures. A current which does not vary with time may exceptionally be obtained owing to the fact that the active material is being formed by the heating at the same rate as the electric field removes it.

These conclusions are strengthened when the decay curves at intermediate temperatures are considered. These have been investigated by Sheard[1] who found that they contained humps similar to those observed by the writer in a good vacuum (p. 199). At temperatures below 628° C. Sheard found that the decay was inappreciable with the platinum used by him. The results at temperatures between this and 774° C. are shown in Fig. 27. Similar curves which showed more pronounced maxima were obtained when the revived emission due to heating in a Bunsen flame was examined at about the same temperatures. These curves can be accounted for if we suppose that three substances are concerned in the emission. One of these A decays continuously with pronounced emission of ions. The second B is formed by the heating and is inactive or comparatively so. B then decays into C with a further positive emission. It will be seen that the curves bear some resemblance to those shown by the decay of the radio-active deposit from radium emanation, where the successive changes have been explained in a somewhat similar manner. The inset represents the radio-active case in which an active product

---

[1] " Phil. Mag.," Vol. XXVIII, p. 170 (1914) ; cf. also Sheard and Woodbury. " Phys. Rev.," Vol. II, p. 288 (1913).

A changes into an inactive product B, from which the active body C is subsequently formed. In the case now under discussion it is not necessary to suppose that B is formed from A, and the phenomena are complicated by the fact that all the rates of change are functions of the temperature ; so that a slight change of temperature may make a considerable difference in the appearance of the curves.

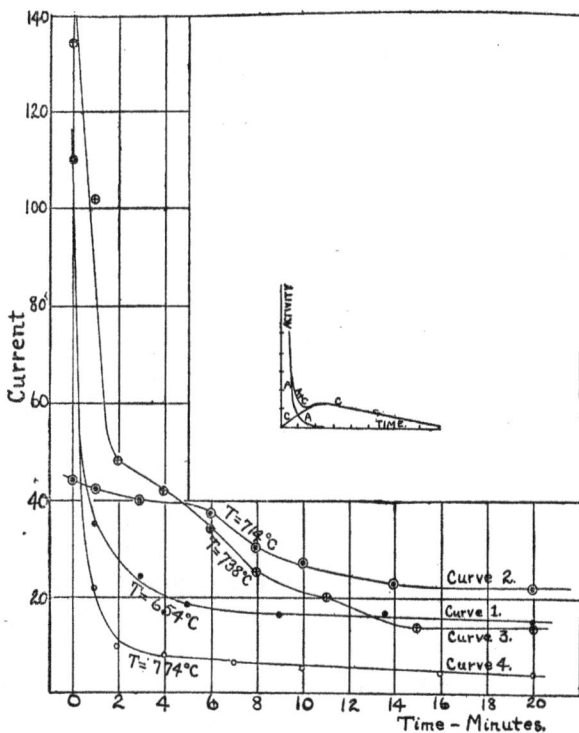

FIG. 27.

The curves connecting current and electromotive-force for the positive currents from freshly heated platinum wires in air exhibit complications similar to those shown by the positive emission from new wires in a high vacuum. At low temperatures the currents may show no indication of approach to saturation, even when the positive currents are quite small, when the negative emission is negligible, and when the time rate of decay of the positive currents also is inappreciable.

Thus with a platinum tube 0·2 cm. in diameter heated to 809° C. and surrounded, in air at atmospheric pressure, by a cold tube 3·2 cm. in diameter, the writer[1] found the relation between the positive currents and the potential difference to be given by the following numbers :—

Volts on hot tube + .	.	.	.	.	.	0	4	10	20	40	80	400	960
Current ($1 = 1·8 \times 10^{-12}$ amp. per cm.^{2})					.	0	2·6	10	22	32	64	225	390

Using the same arrangement with the hot tube at 1200° C. an increase of the potential difference from 80 to 400 volts increased the current only in the ratio 64 to 75. Thus the difficulty in reaching an approximation to saturation with fresh wires appears to occur only at low temperatures. If it is due to the same cause as the similar effect observed in a vacuum (p. 200) this is important, since it would show that neither effect can be attributed to secondary actions arising from the bombardment of the cold cathode by the positive ions. The kinetic energy of the positive ions at the cathode is negligible at atmospheric pressure. As these peculiar effects have, so far, not been explored very fully, it is perhaps undesirable to lay too much emphasis on the precise interpretation of the observed phenomena, but it is difficult to avoid the conclusion that with fresh wires at low temperatures some, at least of the positive ions are in some way liberated at the surface of the wire by the direct intervention of the electric field.

There are distinct indications that under other conditions the electric field may tend to inhibit the formation of the positive ions. Thus at 706° C. in oxygen at 528 mm. pressure the writer[2] found the relation between the positive current and the mean voltage on the filament to be that given by the numbers in the next table. The experiments were made with a thin platinum wire and the readings were taken in the order of the successive columns from left to right :—
Mean volts + :—

0	1·75	38	1·75	3·7	1·75	5·8	18	1·75	38	0

Current ($1 = 6 \times 10^{-13}$ amp.) :—

4	20	14·8	18·5	15·8	19·5	15·5	15	20	14·8	4·8

---

[1] " Phil. Trans., A.," Vol. CCVII, p. 58 (1906).

[2] Loc. cit., p. 7 ; cf. also p. 11.

The current with 38 volts is only about 75 per cent. of that with 1·75 volts. Similar results were obtained at 0·4 mm. and 826° C. and at 0·0015 mm.; so that the gas does not appear to have much to do with this effect. The currents in the table were not those obtained when the changed potential was first applied, but the steady values reached after a few minutes. On raising the potential the currents were larger, and on lowering it smaller, at first. So far as the writer's experience goes this type of behaviour is shown neither by an absolutely fresh wire nor by a well-aged wire but only in the intermediate stages. The results indicate the presence of a substance removable by the electric field which is capable of giving rise to more ions if left in the neighbourhood of the hot metal for some time. Such a state of things might conceivably arise in the stage where the emission increases with lapse of time under otherwise constant conditions.

The maxima sometimes observed in the time decay curves, as well as Sheard's experiments on the revival of the emission at a low temperature by heating in the absence of electric field to various higher temperatures, show that there are at least two distinct substances or actions concerned in the emission of positive ions from freshly heated platinum wires. This conclusion has been confirmed in a different way by Sheard and Woodbury.[1] They heated a fresh wire in air at atmospheric pressure at various temperatures under conditions such that the decay of the emission was inappreciable. The emission was then found to follow the equation $i = AT^{\frac{1}{2}}e^{-b/T}$ with a constant value of b over the range tested (845° K. to 1040° K.). The emission was then allowed to decay until a considerable amount of it had been driven off, when the measurements at different increasing temperatures were repeated. On plotting the value of $\log i - \frac{1}{2}\log T$ against T^{-1} the new curve was found to consist, not of one straight line as at first, but of two straight lines inclined at an angle. This indicates that under the condition of greater aging of the wire the emission at the lower temperatures has one value of b and that at the higher temperatures another. The value of b for

[1] " Phys. Rev.," Vol. II, p. 288 (1913).

the higher temperatures was the same as that which covered the whole range of temperature in the original test.

The three lines of investigation referred to show that the positive emission from fresh platinum wires involves, as a rule, the occurrence of at least two distinct substances or processes. The ions emitted by these substances, or during these processes, are not necessarily different. In order to condense the discussion let us suppose that the observed differences are due to different substances. This hypothesis is most strongly supported by the phenomena described in Chapter VI. The two substances might be derived one from the other by decomposition or they might be different compounds of the same basic element; in either of these events the positive ions emitted from them would be expected to be the same. It is true that there is definite evidence of the emission from the purest available platinum of two well-marked types of ion having values of M about 40 and 24 ; but it cannot be considered certain that these ions correspond, respectively, to the quick initial decay and to the slower decay after passing the maximum, or to the corresponding phenomena discovered by Sheard. As the matter has not been accurately investigated from this point of view it is impossible to be quite certain, but an examination of the evidence at present available indicates that all the various phenomena now under consideration should have been present in the early stages when the platinum wires examined by the writer (p. 218) gave no indications of the presence of any ions except those having a value of M in the neighbourhood of 40. It is probable that similar effects would be observable at the stage at which the lighter ions are emitted, but, so far, there does not appear to be any convincing evidence that they have been examined.

Other interesting properties peculiar to freshly heated platinum wires, many of them closely related to those just considered, will be found described in Chapter IV, pp. 118 *et seq.*, and Chapter VI, *passim.*

CHAPTER VIII.

THE EMISSION OF IONS BY HEATED SALTS.

THE first experiments to indicate that heated salts possessed remarkable electrical properties were made by Sir J. J. Thomson,[1] who showed that the conductivity between platinum electrodes in a hot crucible containing air at atmospheric pressure was much increased by the presence of potassium iodide, potassium chloride, ammonium chloride or sodium chloride. At about the same time Arrhenius [2] found that the conductivity of the Bunsen flame was greatly increased by the injection of various salts. The injection of similar salts of the alkali metals in the proportion of their equivalent weights causes a greater increase in the conductivity the more electropositive the basic element and the higher its atomic weight. This is shown by the following numbers for the conductivities caused by equivalent quantities of salt under a potential difference of $5 \cdot 6$ volts : Cs $= 123$, Rb $= 41 \cdot 1$, K $= 21 \cdot 0$, Na $= 3 \cdot 49$, Li $= 1 \cdot 29$, H $= 0 \cdot 75$. These numbers are taken from a paper by Smithells, Dawson, and Wilson.[3] As the electrical phenomena in flames are probably affected by the chemical actions which occur we shall not consider them further in this book. The reader who desires more information on the subject may be referred to " The Electrical Properties of Flames and Incandescent Solids," by H. A. Wilson (University of London Press : 1912), where it is considered at length.

In 1901 H. A. Wilson [4] examined the electrical conductivity caused by spraying salt solutions into the space between

---

[1] " Phil. Mag.," Vol. XXIX, pp. 358, 441 (1890).
[2] " Ann. der Physik," Vol. XLIII, p. 18 (1891).
[3] " Phil. Trans., A.," Vol. CXCIII, p. 108 (1899).
[4] *Ibid.*, Vol. CXCVII, p. 415 (1901).

two hot coaxial platinum cylinders. The arrangement in fact was that already described on p. 226. The currents were found to be very difficult to saturate, but in most cases saturation was attained by the application of about 1000 volts. The relation between the currents and the temperature was very complicated, doubtless owing to the occurrence of chemical reactions between the salts and the water vapour present. At low temperatures the largest currents were given by potassium iodide and were measurable on a galvanometer at 300° C. At temperatures approaching 1400° C. Wilson found that the saturation currents, with all the salts of the alkali metals tested, became independent of the temperature. Under these circumstances the quantity of electricity transported in unit time was the same as that which, according to Faraday's Law, would be associated with the electrolysis of the salt sprayed into the space between the electrodes in the same interval. This result was verified for the following salts: $CsCl$, Cs_2CO_3, RbI, $RbCl$, Rb_2CO_3, KI, KBr, KF, K_2CO_3, NaI, $NaBr$, $NaCl$, Na_2CO_3, LiI, $LiBr$, $LiCl$ and Li_2CO_3. The salts behave as though each metal atom present were capable of once giving rise to a single ion and then played no further part in the electrical phenomena. Why this happens is not altogether obvious. It may be that the positive ions, which there is reason to believe are atoms of the metal that have lost an electron, are absorbed into the interior of the negative electrode, or they may end their career by forming an inactive chemical compound. The available data are insufficient to decide between the relative merits of these and alternative hypotheses which might suggest themselves.

The effect of the presence of various inorganic substances on the leakage of electricity across a parallel plate air condenser at temperatures in the neighbourhood of 300° C. was examined by Beattie.[1] A large number of substances were found to increase the currents, the most marked effects being obtained with the halogen compounds of zinc, and various mixtures which might be expected to give rise to these bodies. These phenomena have since been investigated by Garrett and

[1] "Phil. Mag.," v. Vol. XLVIII, p. 97 (1899); vi. Vol. I, p. 442 (1901).

Willows,[1] Garrett,[2] and Schmidt and Hechler,[3] among others. An idea of the nature of the phenomena may be obtained by considering the following experiment which may be regarded as typical of a number of those made by these authors. Two parallel metal plates are arranged in an oven so that their temperatures may be maintained at various values up to 400° C. The lower plate can be maintained at various positive and negative potentials whilst the upper, which is insulated, can be connected to an electrometer. The small currents with no salt between the plates are first measured, so as to enable them to be allowed for, and then the currents are determined after the salt under test has been sprinkled on the lower plate. With some salts the current flows only when the plate is positively charged, whereas others cause a leakage of electricity of both signs but usually to different extents. The following list, compiled from papers by Garrett[4] and Schmidt,[5] embraces the substances which have been found to give rise to a considerable amount of ionization at temperatures of about 400° C. :—

Fe_2Cl_6 : Al_2Cl_6 : NH_4Cl : $MgCl_2$: $SnCl_2 + 2H_2O$: $MnCl_2$: $CdCl_2\ddagger$: $ZnCl_2$: CaF_2 : $Al_2F_6\ddagger$: NH_4Br : $ZnBr_2$: $CdBr_2$: NH_4I : CdI_2 : ZnI_2 : NH_4NO_3 : $Cd(NO_3)_2\ddagger$: $Co(NO_3)_2\ddagger$: Quinine sulphate. The substances marked thus $\ddagger$ only caused a leakage when the plate was charged positively. With all the others some effect was obtained with charges of either sign.

The following substances have been found to give little or no ionization at these low temperatures :—

Sn : Pb : Bi : As : Hg : I_2 : $CuCl_2$: $SrCl_2$: $BaCl_2$: LiCl : KCl : $SbCl_3$: $SnCl_4$: $HgCl_2$: Hg_2Cl_2 : KBr : $HgBr_2$: KI : AgI : PbI_2 : HgI_2 : NaF : CuO : ZnO : SnO_2 : Fe_2O_3 : CaO : MgO : $ZnSO_4$: $FeSO_4$: $CuSO_4$: $MgSO_4$: $MgCO_3$: $ZnCO_3$: K_2CO_3 : Na_2CO_3 : $NaHCO_3$: $Pb(NO_3)_2$: $Ba(NO_3)_2$: CH_3OH : C_2H_5OH : $(CH_3)_2CO$: $(C_2H_5)_2O$: $CHCl_3$: C_6H_6 : C_6H_{14} : CS_2 : CH_3COOH : lactic acid : quinone : hydroquinone : naphthalene : phenanthrene : fluorene.

[1] " Phil. Mag.," Vol. VIII, p. 437 (1904).
[2] *Ibid.*, Vol. XIII, p. 728 (1907).
[3] " Verh. der Deutsch. Physik. Ges.," Vol. IX, p. 39 (1907).
[4] " Phil. Mag.," Vol. XIII, p. 729 (1907).
[5] " Ann. der Phys.," Vol. XXXV, p. 404 (1911).

Many of the salts enumerated in this table give a very large ionization at higher temperatures. The behaviour of KI and that of iodine call for special comment. These are given as inactive in the table, whereas Wilson (p. 252) obtained large currents when potassium iodide was sprayed into hot air at about 300° C., and Campetti [1] and Sheard [2] have obtained very considerable currents from iodine vapour at about 400° C. There seems to be little doubt that the currents obtained by Wilson were conditioned by an action between the potassium iodide and the water vapour present. Kalandyk [3] has recently found that the conductivity of KI vapour at 308° C. is negligible, but that it becomes appreciable when water vapour is also present. Why the results of Schmidt's observations with iodine do not agree with those of Campetti and of Sheard is uncertain.

With the type of apparatus just described the measured electrical leakage may arise in a good many ways. It may be caused by an emission of ions of either sign from the heated salt directly, it may be due to the volume ionization of the salt vapour, or it may arise from the emission of ions by the action of the salt vapours on the electrodes. When the salted electrode discharges electricity of both signs all of these actions may be occurring. If only one sign is discharged then there can be no volume ionization, but the current may be due either to the emission of ions of the same sign from the hot salt or of the opposite sign from the opposite electrode by the action of the salt vapours. Thus it is impossible to give a very precise interpretation to the effects obtained with the type of apparatus now under discussion.

Sheard,[4] who has examined the emission from cadmium iodide in some detail, has succeeded in unravelling the various factors to a considerable extent. By using an air-cooled electrode for collecting the ions, he was able to eliminate the possibility of the emission of ions by the action of the salt vapour

[1] " Sci. Torino Atti," 40, 1, p. 55 (1904).
[2] " Phil. Mag.," Vol. XXV, p. 381 (1913).
[3] " Roy. Soc. Proc., A.," Vol. XC, p. 638 (1914).
[4] " Phil. Mag.," Vol. XXV, p. 370 (1913).

on the opposite electrode ; and by allowing the vapours from the salt to pass through the plates of a condenser, charged to a difference of potential sufficient to remove all the ions instantaneously present, and then into a second testing vessel, he was able to examine the processes occurring in the vapour without having to deal with complications due to the ions emitted by the salt. In this way he succeeded in showing that there was an emission of ions directly from the hot salt, and an ionization process in the vapour independent of this. Whether the formation of ions from the vapour is entirely a direct volume ionization, or is in part due to interaction between the vapour and the electrodes, is not absolutely certain. Kalandyk [1] found that the currents through the vapour were not altered much when one of the platinum electrodes was covered with spongy platinum, indicating that the surface of the electrodes was not of much importance. Sheard,[2] on the other hand, found that the currents in the vapour, although apparently saturated, varied very much with the direction of the applied potential difference, a result which points to the contrary conclusion.

Sir J. J. Thomson [3] has tested the leakage of electricity from a number of inorganic substances, in air at atmospheric pressure, and at temperatures for the most part considerably higher than those used in the investigations just referred to. He found that the oxides discharged only negative electricity, the chlorides and phosphates only positive. The nitrates tested discharged only positive electricity until they were converted into the oxides, after which only negative electricity was discharged. In every case, except that of lead peroxide, the sign of the charge which leaked away was opposite to that acquired by the salt when rubbed with a pestle in a mortar.

In Thomson's experiments the salts under test were placed on an electrically heated porcelain tube. He found that the phosphates gave larger currents, when charged positively, than the other groups of salts examined, aluminium phosphate being particularly efficient.

---

[1] Loc. cit., p. 644.   [2] *Ibid.*, p. 380.
[3] " Camb. Phil. Proc.," Vol. XIV, p. 105 (1906).

In most of the recent work on the emission of ions from hot salts, the salts have been placed on an electrically heated strip, or wire, of platinum, which formed one electrode. The other electrode has been cold, and arranged so as to surround the first as far as possible. In a large number of cases, there is no current with this arrangement except when the hot salt is positively charged. Under these circumstances we know that there is no volume ionization, and that positive ions only are emitted, either from the salt directly or by the interaction of the salt vapour on the hot electrode. Similar considerations apply if negative electricity alone is discharged. Under these conditions, the observed currents can be assigned definitely to the *emission* of ions either directly from the hot salt or from the hot electrode under the influence of the salt vapour. The number of possible alternative interpretations of the observed effects is, therefore, considerably reduced. These remarks apply also to the experiments of Thomson, whose apparatus was of this general type. We shall now consider the phenomena in greater detail, keeping for the most part to cases in which the effects are due to an emission of ions in the sense just indicated.

RELATION BETWEEN CURRENT AND POTENTIAL DIFFERENCE.

Naturally this depends a good deal on the shape and relative position of the electrodes, the pressure of the surrounding gaseous atmosphere, whether ions of only one sign or ions of both signs are emitted, the presence or absence of volume ionization, and the magnitude of the emission. In H. A. Wilson's experiments with concentric tubes at atmospheric pressure, where volume ionization and large currents were dealt with, large potential differences of the order of 1000 volts were necessary to attain approximate saturation. At low pressures, and where there is only an emission of ions of one sign from one electrode, the current-E.M.F. curves are similar to those given by the ions emitted from hot metals under parallel conditions. Saturation is usually attained the more readily the lower the temperature and the smaller the current.

As a rule, it is rather more difficult to attain saturation with salted than with unsalted electrodes, although sometimes the reverse is the case. Thus in some experiments in which the writer [1] heated a number of salts in a closed platinum tube 2 cm. in diameter, and measured the currents passing to a central cold electrode 1 cm. in diameter, the currents at a number of potential differences and pressures of air before admitting the salts had the values given in the following table :—

Pressure (mm.) →	Volts . →	40	80	120	160	200	240	280	320	360
0·0075	Current . → 1	1·28	1·45	1·57	1·68	1·77	1·90	2·13	2·31	
0·5 (approx.)	Current . → 1	1·25	1·44	1·54	1·61	1·90	2·45	3·47	5·0	
5·5	Current . → 1	1·29	1·53	1·71	1·80	1·83	1·83	2·5	15	

(1 = 10⁻⁷ amp. approx.).

The substantial increases with the higher voltages at 0·5 and 5·5 mm. pressure are undoubtedly due to impact ionization in the gas. The observations with sodium sulphate in the tube gave practically the same variation of current with voltage at similar pressures as that indicated by the numbers in the preceding table for the empty tube. With aluminium phosphate and beryllium sulphate also the curves were similar except that the current increased somewhat more rapidly with rising potential differences between 40 and 200 volts. It is possible but not certain that this increase is due to impact ionization of the salt vapour close to the hot electrode. In that case it should be more marked with the more volatile salts. Roughly speaking, this requirement appears to be satisfied. The numbers found with barite, the mineral form of barium sulphate, after heating for ten hours, are shown in the next table :—

Pressure (mm.)	Volts . →	0	40	80	120	160	200	240	320	400
0·0015	Current . → 0	1	1·025	1·06	1·05	1·10	1·12	1·14	1·19	
0·8 (1 = 10⁻⁷ amp. approx.)	. → 0	1	1·03	1·11	1·13	1·21	1·24	1·42	1·60	
9·4 (1 = 10⁻⁷ amp. approx.)	. → 0	1	1·12	1·18	1·25	1·31	1·35	1·46		

These numbers show a much better approach to saturation even than the empty tube. However, it is to be remembered that the values for the empty tube were observed in the

[1] "Phil. Mag.," Vol. XXII, p. 669 (1911).

earlier stages of the experiment, and the positive ionization from platinum which has got into the condition of an "old" wire is much more easily saturated than that from freshly heated platinum. The ease of attaining saturation with barite in comparison with the other salts may be due to the possibly smaller volatility of the source of ionization with this material.

The foregoing data for the relation between current and potential difference are only to be taken as representative samples. As we have stated already, the results obtained vary considerably with changes in the conditions enumerated at the beginning of this section. Current-E.M.F. curves for a number of salts in different gases at various pressures with different types of electrodes may be found in the following papers: H. A. Wilson, "Phil. Trans., A.," Vol. CXCVII, p. 424 (1901); Garrett and Willows, "Phil. Mag.," Vol. VIII, p. 446 (1904); Garrett, "Phil. Mag.," Vol. XX, p. 588 (1910); G. C. Schmidt, "Ann. der Physik," Vol. XXXV, p. 440 (1911); Horton, "Roy. Soc. Proc., A.," Vol. LXXXVIII, p. 127 (1913); C. Sheard, "Phil. Mag.," Vol. XXV, p. 370 (1913).

For many experiments it is sufficient to know that saturation or approximate saturation can be attained, and to make sure that this object has been accomplished. The time lag effects which are often observed when the applied potential difference is suddenly changed are considered on p. 266 below.

CHANGES WITH TIME.

In general when salts are heated in a vacuum or in a gaseous atmosphere at constant pressure the saturation currents vary in an interesting way with the time, even when the temperature and the applied potential are kept constant. This effect was first noticed by Garrett and Willows[1] in making experiments with zinc iodide. They found that the positive emission from this substance under conditions apparently constant first increased to a maximum and then diminished.

[1] "Phil. Mag.," Vol. VIII, p. 450 (1904).

The currents i after passing the maximum could be expressed as a function of the time t by means of the equation

$$i = Ae^{-\lambda t} \qquad . \qquad . \qquad . \qquad . \qquad (1)$$

where A and λ are constants. This formula is the same as that which often governs the decay of the initial emission from hot metals (p. 197), and can be accounted for in a similar way by assuming that the emission is due to the decomposition of some substance at a rate proportional to the amount of it instantaneously present. Zinc bromide gave similar results, but the rate of decay of the emission was greater than with zinc iodide. In a later paper Garrett[1] returned to the emission from zinc iodide. He found that the emission did not diminish indefinitely, but that a final steady value was approached asymptotically. The part of the emission which varied with time could be represented throughout the whole range, including the initial rise, by the formula

$$i = A(e^{-\lambda_1 t} - e^{-\lambda_2 t})$$

where A, λ_1, and λ_2 are constants. This formula implies the initial formation of an inactive product which subsequently decays with the emission of ions (cf. Rutherford's "Radioactivity," Chapter IX).

The phenomena exhibited by ordinary laboratory specimens of pure aluminium phosphate have been examined in detail by Garrett.[2] Fig. 28 shows the variation of saturation current with time when this substance is heated at about 1200° C. in an atmosphere of carbon dioxide at 0·5 mm. pressure. The upper curve gives the same data as the lower one on an enlarged vertical scale. This curve shows that the quick initial rise and decay is followed by a slower increase from a minimum to a final steady value. The whole curve is represented very accurately by the formula

$$i = A(e^{-\lambda_1 t} - e^{-\lambda_2 t}) + B(1 - e^{-\lambda_3 t}), \qquad . \qquad (3)$$

with A, B, λ_1, λ_2, and λ_3 constants. This formula implies the inactive formation (A, λ_2) of an active product which quickly decays (A, λ_1) together with the independent inactive

---

[1] " Phil. Mag.," Vol. XIII, p. 745 (1907).
[2] *Ibid.*, Vol. XX, p. 577 (1910).

formation of a product (B, λ_3) which decays at an infinitely slow rate (B, $\lambda_4 = 0$) with emission of ions. The values of the constants vary with the temperature; so that the general appearance of the curves changes considerably according to the temperature. When the salt was heated in air or hydrogen the initial rise was preceded by a quick decay from a large initial value. This part of the curve did not appear if the

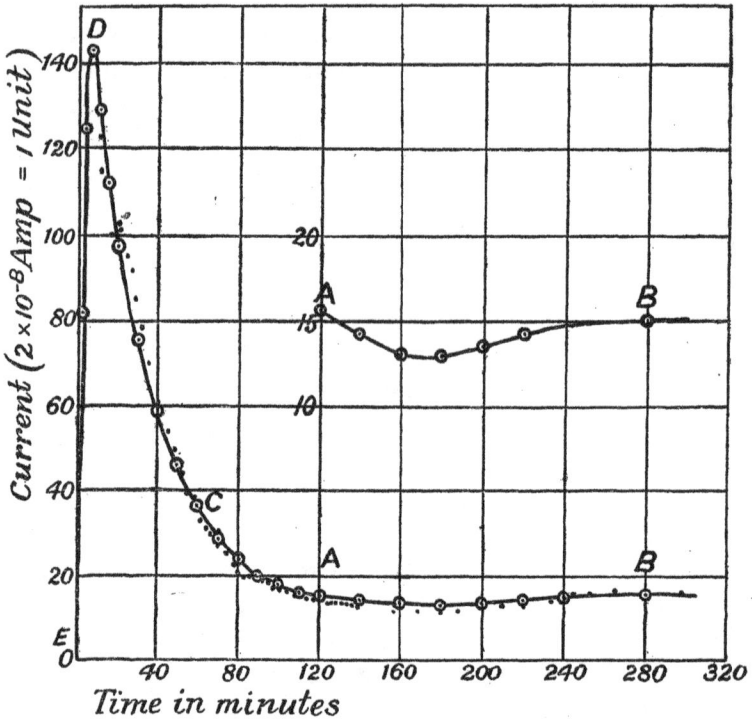

FIG. 28.

salt was previously heated at a lower temperature sufficiently high to drive off observed water vapour; so that, on these and other grounds, it is attributed by Garrett to the action of water vapour.

Time changes of the character under discussion are a general feature of the emission when ordinarily prepared samples of salts are first heated. In addition to those already mentioned a number of cases have been investigated by G. C.

Schmidt.[1] They include ZnI_2 in nitrogen, $AlCl_3$ in nitrogen, $CdCl_2$ in air, $CdBr_2$ in air and CdI_2 in nitrogen. A number of these were examined at different pressures and temperatures. Under the conditions of Schmidt's experiments the emission from the cadmium salts fell away from the beginning and did not show an initial rise to a maximum. Similar observations with Na_2SO_4 in a good vacuum at 1005° C. have been recorded by the writer,[2] and on sodium pyrophosphate and the phosphates of sodium and aluminium by Horton.[3]

Effects of a like character are observed also when negative ions, whether heavy ions or electrons, are emitted by salts (see p. 108). Thus with calcium iodide the writer[4] observed an

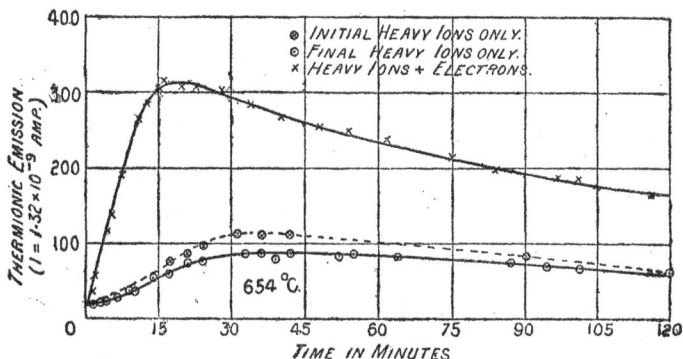

FIG. 29.

initial rise to a maximum in about 15 minutes followed by a slower decay, at temperatures between 523° C. and 654° C. At the higher temperatures the maximum was attained by the electrons sooner than by the heavy ions; at the lower temperatures there was no noticeable difference in this respect. The variation of these currents with time at 654° C. is shown in Fig. 29.

The phenomena exhibited by cadmium iodide have been examined in some detail by Sheard,[5] who tested both the

[1] " Ann. der Physik," Vol. XXXV, p. 401 (1911).

[2] " Phil. Mag.," Vol. XXII, p. 676 (1911).

[3] " Roy. Soc. Proc., A.," Vol. LXXXVIII, p. 134 (1913).

[4] " Phil. Mag.," Vol. XXVI, p. 464 (1913).

[5] *Ibid.*, Vol. XXV, p. 370 (1913).

conductivity of the vapour and the emission of ions from the salt. At temperatures below the melting-point of the salt (400° C.) the saturation currents in the vapour decayed continuously from a maximum initial value, in agreement with Schmidt's results. At higher temperatures there was a rise to a maximum in about 15 minutes followed by a slower decay. The currents due to the emission of ions from the heated salt showed a different behaviour from those in the vapour. At 470° C., for example, there was an enormous negative emission which decayed very rapidly with time. The positive emission was at first too small to measure, but it gradually increased to a maximum value in 90 minutes and then fell away. At this stage the positive emission was greater than the negative, but the greatest positive emission was less than one two-hundredth part of the large negative emission observed on first heating. A similar but less marked contrast between the positive and negative emissions was observed when iodine was similarly tested. Sheard also examined the behaviour of the salt which distilled out of the experimental tube in successive experiments. He found that the first distillate gave a small negative and a large positive emission whereas the second showed the contrary behaviour. In all the distillates there was a great disparity in the magnitudes of the positive and negative emissions ; and in almost every case the distillate from a preparation which gave a large negative and a small positive emission, or vice versa, showed the contrary behaviour. The currents from all the distillates were much smaller than the large initial emission from the fresh salt. The distilled salt showed no appreciable change in appearance, but chemical analysis showed that successive distillation reduced the percentage of iodine.

There can be little doubt that these interesting time changes in the emission of ions from salts and in the conductivity of salt vapours are symptomatic of the occurrence of chemical changes ; but it is very difficult to form a definite opinion as to what the precise nature of the change is, in any particular case. When the currents are increasing with time it seems fairly clear that a substance possessing greater thermionic activity is being formed and when the currents are diminishing the resulting products are less active in this respect. One difficulty in forming a judgment as to the nature of the chemical changes arises from the delicacy of the electrical test. This is so sensitive that the amount of matter concerned might often be incapable of detection by chemical methods. It is also possible that many of the effects are due to the occurrence of unstable forms which are not persistent enough to be recognized by chemical methods. This is especially likely since the time changes show that the bodies concerned have only a transitory existence. In many cases these time changes are attributable to the presence of contaminants. Thus ordinary laboratory specimens of " pure " aluminium phosphate give an initial emission which is large compared with that from the pure salt and which after a time falls to a small value. Horton[1] has shown by spectroscopic examination that this decay in the emission is accompanied by the disappearance of sodium salts.

The complicated phenomena in the case of cadmium iodide have been studied more fully, perhaps, than those shown by any other salt, and here it does seem possible to form, at any rate, a limited judgment as to the nature of the phenomenon from the chemical side. Schmidt[2] has ventured the opinion that the time changes in the vapour arise from the decomposition of the molecules of CdI_2 into $Cd++$ and I_2-- with a subsequent interchange resulting in $Cd + -$ and $I_2 + -$, that is to say, two neutral molecules. It does not seem to the writer,[3] however, that any theory of this type will account for the observed time changes in the vapour in presence of an excess of salt. So long as there is any excess of salt the vapour will be supplied at a steady rate and the phenomena observed in it should be independent of time until the salt disappears. It is necessary to suppose that the actions in the vapour are not conditioned solely by the amount of CdI_2 vapour present but

[1] Loc. cit.

[2] " Ann. der Physik," Vol. XXXV, p. 428 (1911). These views are modified somewhat in a later paper (*ibid.*, Vol. XLI, p. 673 (1913)) without, however, overcoming the difficulty referred to (cf. p. 248).

[3] O. W. Richardson, " Phys. Rev.," Vol. XXXIV, p. 387 (1912).

rather by some other substance coming from the salt. The time changes must in fact be conditioned by something the amount of which is determined by actions occurring at the salt and not simply by a decomposition of cadmium iodide vapour. In one aspect this question has been definitely settled by Kalandyk [1] who has shown that the currents in cadmium iodide vapour under the conditions of these experiments are independent of the time, provided every trace of water is removed from the salt and from the apparatus. The way in which water brings about the time changes usually observed is unknown. Kalandyk's experiments only tell us that there are no time changes when water is absent, they do not offer an explanation of the changes which occur in the presence of water or water vapour. Sheard's results point to the conclusion that the large negative initial emission, when it is present, is connected with the liberation of iodine. On this view the smaller negative emission from the distillates would be related to the reduced iodine content of the salt, which after distillation probably consists of a solution of an unrecognized subiodide of cadmium in CdI_2. The presence of the subiodide would reduce the equilibrium pressure of iodine in presence of cadmium iodide vapour. The probable existence of a subiodide of cadmium is distinctly indicated by the work of Morse and Jones [2] who succeeded in isolating a body having the composition $Cd_{12}I_{23}$, probably a solution of the subiodide in CdI_2.

It is likely that the effect of water vapour is not confined to this particular instance and that many of the time changes observed with other salts would not occur if all traces of water were eliminated. Such a result, at any rate, would not be surprising if the time changes are indicative of the occurrence of chemical reactions. For it is well known that many chemical actions which proceed very energetically in presence of a trace of water vapour are completely inhibited in its absence. The importance of water vapour generally for these effects is supported by the behaviour of potassium iodide, whose vapour

[1] Loc. cit.
[2] " Amer. Chem. Jour.," Vol. XII, p. 488 (1890).

exhibits little or no ionization at low temperatures if water vapour is completely absent. Again, as we shall see later, there is a close correspondence between the emission of posi- tive ions from salts and from fresh platinum wires, and W. Wilson[1] has found that the positive emission from the latter is, under certain conditions, very sensitive to the presence of small quantities of water.

In a recent paper Schmidt[1] has come to the conclusion that the time changes previously observed by himself and others are to be attributed entirely either to removal of ions by the electric field or to diminution of the salt surface, in the case of a decrease of ionization with time, or to a time lag in the temperature or pressure of the vapour reaching the elec- trodes, in the case of an increase with time. The main grounds for this conclusion are (1) that the currents are greater when the same amount of the salt is tested in the powdered form as compared with a pastille, indicating that the amount of surface is a factor, and (2) when conditions are arranged so that the superficial area of the salt does not change during an experiment, the time variations disappear. Although the first of these grounds is probably correct it does not seem to the writer that either of them is established by the experi- ments described by Schmidt. In these experiments all the currents are measured under a potential difference of only 2 volts, and they must have been very far from saturation. It is well known that under such conditions the magnitudes of the currents may be almost independent of the number of ions available for carrying them, the main factor in determining their values being the mobilities of the ions. Schmidt's con- clusions are also in direct contradiction to the experimental results of Sheard, who undoubtedly observed in the same tube a simultaneous decrease in the negative and increase in the positive saturation currents, both effects changing at different and characteristic rates.

The time changes we have had under consideration so far are such as arise when a constant potential difference is main- tained between the electrodes. In many cases this decay

[1] "Ann. der Physik," Vol. XLI, p. 673 (1913)

appears to be due merely to heating and to be independent of the magnitude or sign of the electric field. This is not, however, a universal rule. With aluminium phosphate the writer has observed that the general decay of the positive emission with time is much more marked when the salt is positively charged than when it is unchanged or negatively charged. The phenomenon has, however, not received much attention. We have already remarked upon similar effects exhibited by the positive emission from fresh wires (pp. 200 and 244).

Apart from this it has generally been found that immediately after changing the sign of the applied potential difference the currents of either sign are larger than the relatively steady values to which they shortly settle down. Effects of this kind have been recorded by H. A. Wilson[1] with salts of the alkali metals heated in air at atmospheric pressure, by Schmidt[2] with zinc and cadmium iodides in various gases at low pressures, and by Garrett[3] and by the writer[4] with aluminium phosphate in a vacuum.[5] As a rule these changes are soon over and are independent of the general decay or increase with time already considered. This is well shown in the case of zinc and cadmium iodides by the curves given by Schmidt. In the case of the specially prepared pure aluminium phosphate referred to below, the writer found that the effect of changing the electric field was smallest at low and high temperatures and most marked at intermediate temperatures. These effects thus appear to depend to some extent on the temperature of the salt.

With some salts when the temperature is suddenly changed the emission assumes an abnormal value for a short time. Thus when sodium phosphate had been overheated Horton[6] observed that the currents at a given lower temperature were abnormally high at first. A similar effect has been noticed

[1] " Phil. Trans., A.," Vol. CXCVII, p. 415 (1901).
[2] " Ann. der Physik," Vol. XXXV, p. 428 (1911).
[3] " Phil. Mag.," Vol. XX, p. 577 (1910).
[4] *Ibid.*, Vol. XXII, p. 700 (1911).
[5] Cf. also Horton, " Roy. Soc. Proc., A.," Vol. LXXXVIII, p. 117 (1913).
[6] " Camb. Phil. Proc.," Vol. XVI, p. 92 (1910).

by the writer[1] with sodium sulphate. In the case of salts like calcium iodide which emit a mixture of electrons and heavy ions the writer[2] has observed a time lag in the current caused by changing an external magnetic field. In fact a sudden change in any physical condition controlling the magnitude of the thermionic current appears temporarily to upset the internal conditions which determine the value of the saturation current under given external conditions.

In the case of aluminium phosphate Horton[3] has observed a decay in the steady emission when the salt is left in air at a low pressure in the cold. The writer is inclined to suspect that this effect is connected with the gradual dehydration of the salt, but there is not enough evidence to form a certain judgment on the point. Similar effects have been observed by the writer in the case of the negative emission from calcium iodide (p. 108).

VARIATION WITH TEMPERATURE.

The ionization currents from salts or in salt vapours as ordinarily measured may exhibit very complicated changes when the temperature is varied. Thus H. A. Wilson in the experiments already described, in which salts were sprayed into the hot air between two co-axial platinum cylinders, found that the curves expressing the relation between current and temperature possessed maxima and minima at certain temperatures. These complications are undoubtedly due to the occurrence of chemical reactions in such a way that the ionization is caused by different substances at different temperatures. The particular effects observed by Wilson were probably caused by the formation of hydrates owing to the action of the salts on the water vapour present. The chemical actions, whose precise nature is less obvious, which give rise to the time changes considered in the preceding section probably cause the complications which are frequently observed in the relation between emission and temperature in other cases when salts are heated. It is clear that the frequent occurrence of chemical action greatly increases the difficulty of interpreting

[1] "Phil. Mag.," Vol. XXII, p. 680 (1911).   [2] *Ibid.*, Vol. XXVI, p. 465 (1913).
[3] "Roy. Soc. Proc., A.," Vol. LXXXVIII, p. 126 (1913).

experiments designed to discover the relation between emission and temperature when a given salt is heated.

In spite of these difficulties there is a very considerable amount of experimental evidence which goes to show that when a salt is heated under conditions such that the emission of ions is always caused by the same substance the currents increase rapidly and continuously with rising temperature, and the relation between the total emission (or the saturation current) and the temperature is that expressed by the formula

$$i = AT^{\frac{1}{2}}e^{-b/T}$$

which has been found to govern the temperature relations of other thermionic currents. Thus Garrett[1] showed that this relation held when a number of salts were heated on a brass plate at temperatures ranging around 300° C. Data which lead to a similar conclusion have been furnished by Garrett[2] for the positive emission from aluminium phosphate in CO_2 and H_2 at 0·05 mm. pressure at about 1100° C., by Schmidt[3] for cadmium iodide, by the writer[4] for the negative emission consisting of a mixture of electrons and heavy ions which is given off by calcium iodide, strontium iodide, and calcium fluoride, and by Kalandyk[5] for the currents through the vapours of cadmium iodide, zinc iodide, and zinc bromide. The values of the constant b deduced from some of these experiments are shown in the following table:—

Nature of Emission.	Pressure.	Substance.	Authority.	Approximate Mean Temperature.	Value of b° C.
Positive	Atmospheric	CaF_2	Garrett	297° C.	1·3 × 10⁴
,,	,,	AlF_3	,,	330° C.	1·45 × 10⁴
,,	,,	NH_4NO_3	,,	312° C.	1·65 × 10⁴
Negative	,,	ZnI_2	,,	241° C.	1·45 × 10⁴
,,	,,	$FeCl_3$	,,	355° C.	3·05 × 10⁴
,,	,,	NH_4Cl	,,	352° C.	2·5 × 10⁴
,,	,,	CaF_2	,,	346° C.	3·0 × 10⁴
,,	,,	NH_4NO_3	,,	342° C.	2·15 × 10⁴
,,	,,	$MgCl_2$	,,	326° C.	1·2 × 10⁴
Positive	0·05 mm. of CO_2	Aluminium Phosphate	,,	1100° C.	3·55 × 10⁴
,,	,, ,, H_2	,,	,,	1200° C.	2·65 × 10⁴

[1] "Phil. Mag.," Vol. XIII, p. 732 (1907).
[2] *Ibid.*, Vol. XX, p. 581 (1910).
[3] "Ann. der Physik," Vol. XXXV, p. 401 (1911).
[4] "Phil. Mag.," Vol. XXVI, p. 452 (1913).
[5] "Roy. Soc. Proc., A.," Vol. XC, p. 642 (1914).

Nature of Emission.	Pressure.	Substance.	Authority.	Approximate Mean Temperature.	Value of b° C.
Negative	0·001 mm.	CaI_2	Richardson	500° C.	2·76 × 10⁴
,,	,,	SrI_2	,,	540° C.	5·76 × 10⁴
,,	,,	CaF_2	,,	600° C.	3·64 × 10⁴
Current through vapour	Constant	CdI_2	Kalandyk	250° C.	2·35 × 10⁴
,,	,,	,,	,,	400° C.	1·14 × 10⁴
,,	,,	$ZnBr_2$	,,	400° C.	1·74 × 10⁴
,,	,,	ZnI_2	,,	380° C.	1·57 × 10⁴

With the positive emission from zinc iodide in air at 2·5 mm. pressure Garrett found that there was a break in the curve obtained by plotting log $iT^{-\frac{1}{2}}$ against T^{-1} at 250° C.; this was attributed to a fresh source of ions coming into play at this temperature.

The values of b given in the table above are all of the same order as those given by the emission of ions from hot metals. For the most part they tend to run lower than the values characteristic of the negative emission from most metals and are more comparable with the values for the positive emissions. The writer found that there was no certain difference, at any rate over considerable ranges of temperature, in the values of b for the heavy ions and for the electrons, in the case of the three salts CaI_2, SrI_2, and CaF_2 which give off a mixture of these bodies. Kalandyk's experiments were made in such a way as to vary the temperature of the salt vapour without changing that of the salt. Thus the pressure of the salt vapour in these experiments was presumably constant and equal to the vapour pressure of the salt at the temperature at which the latter was maintained. It does not seem likely that the difference between the two values of b for CdI_2 given by Kalandyk is due to the difference of mean temperature merely; but the matter has not been sufficiently investigated to enable the precise cause of this difference to be ascertained. In considering his results Kalandyk uses the formula $Ae^{-b/T}$ instead of $AT^{\frac{1}{2}}e^{-b/T}$. This of course alters the values of the constants somewhat; except for this, there is no detectable difference between the behaviour of the two functions over the range of T covered by the experiments; so that for the purpose of expressing the numerical values there is nothing to choose between these formulæ.

It is to be remembered that the values of b given in the table can only be relied upon as being approximately correct, as in many cases the conditions other than temperature which affect the formation of the ions have been insufficiently investigated.

THE INFLUENCE OF THE NATURE AND PRESSURE OF THE SURROUNDING GAS ON THE THERMIONIC CURRENTS FROM SALTS.

Garrett[1] observed that the currents from aluminium phosphate, when positively charged and heated on a strip of platinum at a constant temperature, varied in a regular manner with the pressure of the surrounding gas. This effect occurs when the currents are approximately saturated; so that it must be caused by a change in the actual number of ions emitted, and cannot be due merely to a change in the mobility of the ions. The effect of varying the pressure and keeping the other conditions constant was found to be as follows :—

At the lowest pressures the emission of ions was quite small, but it increased steadily with rising pressure until a certain pressure was reached at which the emission had a maximum value. After this the emission diminished at a rate which was smaller than that of the previous rise, and which fell off continuously as the pressure was increased. Although the diminution of the currents with rising pressure fell off as the pressure increased, it was still quite noticeable at about 50 mm. pressure in the neighbourhood of 1100° C. Results of much the same character were obtained in both air and carbon dioxide. The pressure at which the current attained a maximum value was found to diminish very considerably as the temperature of the salt was raised.

This passing of the current through a maximum value as the pressure is raised, is similar to the behaviour of an ionized gas subject to a constant potential difference, in the range of pressure in which a large part of the current is due to impact ionization. It does not appear, however, that impact ionization can be the cause of the phenomena now under considera-

[1] "Phil. Mag.," Vol. XX, p. 579 (1910).

tion. In the first place, they occur under conditions such that the currents vary very little with moderate changes in the applied voltage. In the second place, they occur with voltages which are not large enough to give rise to any appreciable amount of impact ionization. Finally, the change with temperature, of the pressure for maximum current, is in the wrong direction, and also the rate of change is too great, to be in agreement with this explanation.

The phenomenon in question is not confined to aluminium phosphate. Similar observations have been made by Horton[1] on sodium and lithium phosphates, and by the writer[2] on Na_3PO_4 and Na_2SO_4. Horton's experiments with sodium phosphate were made at 800° C., and the effect of the different gases, carbon monoxide, hydrogen, and oxygen was examined. The largest emission was observed in hydrogen, but it decayed more rapidly with time than that in the other gases. The emission in carbon monoxide was about ten times as large as that in oxygen, although the curves connecting emission and pressure were similar. The maximum in hydrogen was not detected, as the currents increased continuously up to the highest pressure (20 mm.) at which experiments were made. In a later paper Horton[3] showed that the behaviour of lithium phosphate was similar to that of sodium phosphate in these respects. The writer's observations on Na_3PO_4 and Na_2SO_4 in air showed that under comparable conditions these substances behaved in much the same way as sodium phosphate in oxygen, as recorded by Horton. With Na_2SO_4, which was examined over the range of temperature from 730° C. to 1160° C., the following additional points were noted, among others. The maximum emission at a certain pressure which was observed at the lower temperatures (about 800° C.) with relatively fresh salt, was found to disappear if the salt had been heated for a long time at a high temperature (about 1150° C.) before testing. The maximum at the higher temperatures was not observed to disappear under this treatment.

[1] "Camb. Phil. Proc.," Vol. XVI, p. 89 (1910).
[2] "Phil. Mag.," Vol. XXII, p. 676 (1911).
[3] 'Camb. Phil. Proc.," Vol. XVI, p. 318 (1911).

The effect of water vapour, instead of air, was also tried. The emission in water vapour was about six times as great as in air, and the pressure of maximum emission was found to be raised from 0·2 mm. to 0·5 mm. at 1160° C. The emission in air was subsequently found to have been permanently diminished by the treatment with water vapour. The variation of the emission in air with pressure at the lowest pressures was carefully tested. The effect of the gas was found to be very irregular. Sometimes the magnitude of the emission would be very sensitive to the admission of a small amount of air and at other times very insensitive under conditions apparently identical. In all cases the relation between current and pressure was of the form $a + bp$, where a and b are constants, provided p was sufficiently small.

Horton [1] has since extended his observations to 1080° C. and 1190° C. and has examined sodium pyro-phosphate and several specimens of aluminium phosphate as well as sodium ortho-phosphate, with results for the most part similar to those already described. Two specially pure specimens of aluminium phosphate prepared by the method indicated on p. 294 failed to exhibit the pressure of maximum emission. The emission from these preparations is very small and may possibly be due to the underlying platinum. Similar results were obtained with the small emission from the impure aluminium phosphates which had been heated for a long time. With this salt the maximum, as well as most of the emission, is clearly due to some impurity which disappears with continued heating. The magnitudes of some of the positive emissions at various pressures of air, which were obtained after continued heating, are shown in the accompanying table :—

POSITIVE THERMIONIC CURRENTS ($i = 10^{-9}$ AMP.) IN AIR AT VARIOUS PRESSURES AT 1190° C.

Material.	0·005 mm.	1 mm.	2 mm.	5 mm.	10 mm.	20 mm.
Platinum	5·2	1·65	1·65	2·2	2·9	3·9
Pure aluminium phosphate .	0·7	0·6	0·7	0·9	1·2	2·0
Impure aluminium phosphate .	2·9	1·5	2·0	3·1	4·1	5·0
Sodium phosphate . .	1080	1500	1630	1750	1810	1740

[1] "Roy. Soc. Proc., A.," Vol. LXXXVIII, p. 117 (1913).

The initial emission from the impure aluminium phosphate is larger than that for the sodium phosphates at these temperatures, and like them it shows the pressure maximum. It also decays at an enormously more rapid rate. In these experiments an increase of current with diminishing pressures was observed at very low pressures, in most cases. This is attributed by Horton to an action between the heated anode and the mercury vapour, but it seems possible that the alterations of pressure in this region may have caused a difference of temperature between the thermocouple and the emitting surface; so that with a constant thermocouple reading the temperature of the hot surface may vary with the pressure of the gas.

In all these experiments small quantities of salt were used and the salts were heated electrically on a strip of platinum. In order to vary the conditions as much as possible the writer[1] made experiments in which a number of salts were heated at the bottom of a long platinum test tube. To prevent the platinum tube from collapsing it was placed in an exhausted steel crucible heated in an electric furnace. The currents from the platinum tube to an air-cooled central electrode were measured. They were approximately saturated. The salts tested were: Na_2SO_4, $BeSO_4$, $AlPO_4$, and $BaSO_4$. In the last case both the chemically prepared salt and the mineral barite were used. With this apparatus the relation between the saturation currents and the pressure of the air in the tube was found to be quite different both for the different salts as compared with each other and also as compared with the same salt when tested by the strip method. The nature of these differences is illustrated by Figs. 30 and 31. Curve 1 in Fig. 30 shows the variation of positive emission with pressure of oxygen for sodium phosphate as observed by Horton by the strip method at 800° C. Curve 2, Fig. 30, shows the behaviour of Na_3PO_4 by the tube method in air at 775° C. The effect of changing the pressure in the one case is almost the exact opposite of what it is in the other. Separate experiments have shown that this difference cannot be

[1] Loc. cit.

18

attributed to the difference of temperature or of the gases
used in the two experiments.   In Fig. 31 similar observations

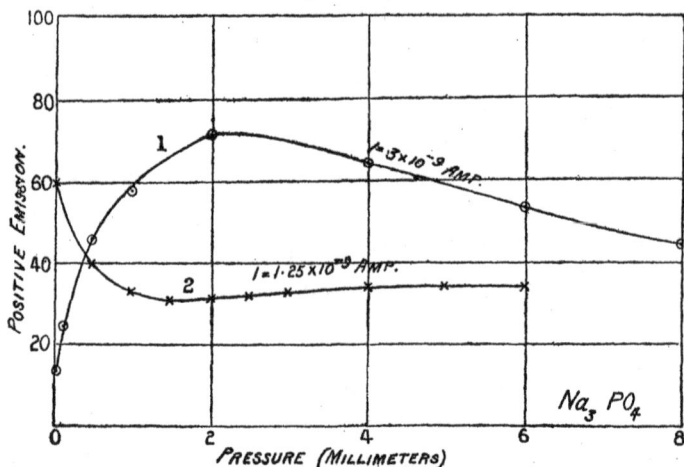

FIG. 30.

with aluminium phosphate are exhibited.   Curves 1 and 2
show the results obtained when ordinary " chemically pure "
aluminium phosphate is tested in air by the tube method at

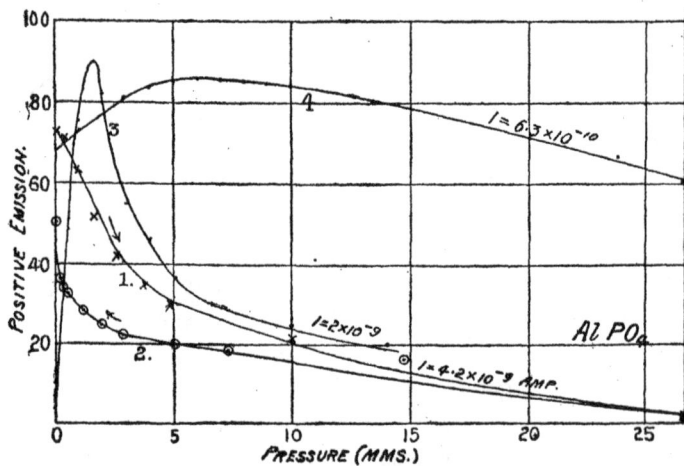

FIG. 31.

780° C.   Curve 1 gives the observations for rising and curve 2
for diminishing pressures.   Most of the difference between
these curves is due to a time lag in the effect of changing the

pressure, but part of it is due to a drift in the temperature of the tube. The mean of the two curves can be taken to represent the actual effect of pressure at the mean temperature (780° C.). Curve 3 shows the quite different results obtained with aluminium phosphate in carbon dioxide by Garrett at 1005° C. by the strip method. According to Garrett's experiments the only effect of reducing the temperature from 1005° C. to 780° C. would have been to shift the maximum towards higher pressures, apart, of course, from the inevitable reduction of the value of the current at any pressure with reduced temperature. Thus, as similar results were obtained by Garrett in air, in this case also the difference between the curves cannot be attributed to the difference in the gases and temperatures used. Curve 4 shows some observations with specially prepared pure aluminium phosphate in the tube. As the current from this substance was only of the same order as that given by the empty tube it is perhaps questionable whether the observed effects were really caused by the salt. The pressure-emission curves with barite were quite different from those given by the chemically prepared $BaSO_4$. In fact, as tested by the tube method, the salts Na_2SO_4, Na_3PO_4, $BeSO_4$, $BaSO_4$, barite, and the two specimens of aluminium phosphate all gave rise to curves which were quite different one from another. For a fuller account of these differences the original paper must be consulted.

The behaviour of barite, which was examined in some detail, exhibited a number of points of interest. In contrast with most other salts the emission from this substance appeared to increase with continued heating in a vacuum. The original small value could be restored by heating the salt in air at atmospheric pressure. This points to the conclusion that the increased emission with continued heating is due to the formation of reduction products. This conclusion is strengthened by the fact that the tubes usually smelt of sulphurated hydrogen after carrying out a test at low pressures, and by the fact that the emission from the salt was found to be increased after it had been heated in hydrogen. Still larger currents, however, were obtained during the heating in hydrogen, when the process of reduction was in active operation. There is thus distinct evidence here of an emission of ions caused by chemical action. When the salt had been heated in air at atmospheric pressure, and the pressure was changed so rapidly that there was little chance of any alteration in the composition of the salt taking place, the emission was practically independent of the pressure of the air from 760 to 0·002 mm. Small changes in the emission were actually observed in carrying out such an experiment; but there are a number of subsidiary causes which might fully account for them, and, in any event, the changes which were observed were negligible compared with those which occur when salts heated on a platinum strip are treated similarly.

It is clear from these results that the emission of ions from salts cannot be regarded as a function of the pressure of the surrounding gas merely, at any rate without further specification. The most striking differences between the results of the experiments with the tube as compared with the strips are: (1) The very varied individual behaviour of the salts when tested by the tube method. With the strip method these differences disappear and are replaced by a definite type of curve with one maximum. This behaviour is shown by all the salts and all the gases which have been tested. On account of the very varied chemical characteristics of the gases used, this uniformity points to a physical and not to a chemical phenomenon, so far as the action between the gas and the salt influences the emission. This physical effect of the gas must be one which is present when the salt is heated by the strip method but not when the tube method is used. (2) With the strips the emission is very sensitive to a small increase of gas pressure at low pressures. In the tube experiments this sensitiveness is not observed. With some salts the emission increases a little with rising pressure, with others it diminishes.

A large number of the facts can be brought into agreement if one assumes that the emission is conditioned partly or entirely by an interaction between the hot electrode and a vapour given off by the salt. At a low pressure, in the strip experiments, such a vapour would easily diffuse away from the hot electrode. There would be no corresponding opportunity with the heated tube. If gas were admitted in the strip experiments this diffusion would be prevented and the vapour would be thrown back on to the strip; so that up to a certain point there would be observed a rapid increase of emission with rising pressure. The cause of the falling off in the emission at higher pressures is less clearly indicated. It seems most likely to arise from the cooling of the salt surface by the gas; so that the temperature of the surface of the salt diminishes with rising pressure when the temperature of the underlying strip is kept constant. In this way the amount of vapour available for the process which causes the emission of ions would fall off as the pressure rises. There are a number of other causes which might give rise to a similar effect, so that it is, perhaps, undesirable to lay too much stress on this particular explanation.

The foregoing explanation of the increase of current with gas pressure observed in the strip experiments is confirmed by the occurrence of a phenomenon which was frequently noticed in the tube experiments at low pressures. On letting in more gas the immediate response of the emission was always in the direction of lower values followed by a gradual adjustment to the steady value characterizing the new pressure. On diminishing the pressure a similar, but less marked, set of changes in the contrary direction was observed. On the explanation referred to, the immediate effect of letting in more gas would be to compress the vapours into the bottom of the tube, reduce the amount of vapour in contact with the hot platinum, and so diminish the current. The subsequent recovery would be due to the diffusion of the vapours into the fresh gas. The contrary effect on reducing the pressure may be accounted for on similar lines.

There is one very important point which has not, so far, been mentioned in discussing these effects. We shall see in the next section that the positive ions emitted from heated salts, even in an atmosphere of gas, appear to consist of charged atoms of some metal present in the salts. There is no indication of the occurrence of positive ions whose electric atomic weights have values corresponding to those of the atoms or molecules of the surrounding gases, at any rate as a general feature of the phenomena. Thus the effect of gases on the emission of ions from salts must be an indirect one. It is not a process involving ionization of the gas.

As regards the very varied curves given by the different salts when tested by the tube method, all that it seems desirable to say at present is that they are probably symptomatic of the chemical changes occurring. The emission at any pressure must depend on the chemical composition of the salts and salt vapours present. This is changed by altering the pressure of the gas, and the emission follows the pressure changes in a corresponding way. The complexity of the curves is to be expected, as the reactions are known to be very involved.

Some peculiar phenomena displayed by the negative emission from calcium iodide (cf. Chap. III, p. 107) which were observed by the writer[1] after this salt had been allowed to stand in the cold in air and in a vacuum may possibly be related to those just discussed.

SPECIFIC CHARGE (ϵ/m) AND ELECTRIC ATOMIC WEIGHT, (M) OF THE IONS.

The first experiments to measure ϵ/m for the positive ions from salts were made by Garrett[2] with aluminium phosphate using the method due to Thomson which is described on p. 8. From these experiments Garrett concluded that about 10 per cent of the ions emitted had an electric atomic weight equal to or less than that of hydrogen. This conclusion has not been confirmed by experiments with aluminium phosphate made by Davisson[3] by another method (see below, p. 285). In examining other salts also, the writer has frequently looked for evidence of the presence of ions having values of M of this order of magnitude without finding any. Although such ions may, for some reason at present unknown, have carried part of the current under the particular conditions of Garrett's experi-

---

[1] "Phil. Mag.," Vol. XXVI, p. 452 (1913).
[2] *Ibid.*, Vol. XX, p. 582 (1910).   [3] *Ibid.*, p. 139 (1910).

ment, it seems quite certain that they do not play an important rôle, as a rule, in the emission of positive ions either from aluminium phosphate in particular or from salts in general.

Measurements of the electric atomic weights of the positive ions from the salts of the alkali metals have been made by the writer,[1] using the slit method described on p. 212. In the first instance the sulphates of all the alkali metals, lithium, sodium, potassium, rubidium, and cæsium were examined. We have seen that the values of ϵ/m and M are determined by the horizontal displacements, between the maximum points in the curves, due to a reversal of the deflecting magnetic field. These curves represent the proportion of the total number of emitted ions which pass through the slit for different horizontal displacements of the latter. In general, with specimens of salt which are ordinarily regarded as "chemically" pure, it is found that the nature of the displaced curves and the distance between the maxima depends on the time during which the salt has been heated. This is well illustrated by the observed behaviour of lithium sulphate. On first heating, the magnetically displaced curves had a single maximum at the position corresponding to M = 35·9. After heating for 12 hours each displaced curve possessed two maxima, although there was only one in the undisplaced curve. This shows that two kinds of ions were present which were deflected to different extents by the magnetic field. The value of M for the least deflected was found to be 41·8, whilst the outside maxima gave M = 5·5. As the heating was continued the positions of these maxima remained practically unchanged, but the inner maxima became smaller and smaller with continued heating. After 44 hours the inside maxima were inappreciable and the outer maxima gave M = 5·57. After 52 hours the conditions were much the same and a measurement of M yielded the value 7·43. With further heating the outer maxima disappeared gradually as the salt volatilized and a new inner maximum appeared. After 70 hours' heating this maximum was at the position corresponding to M = 20·6. At this stage the currents are small and the emission has a

[1] "Phil. Mag.," Vol. XX, pp. 981, 999 (1910).

value of M very near to that from a platinum strip which has been heated for a long time (cf. p. 218).

In this experiment, although the ions which were first emitted had a value of M near 40, the greater part of the whole number of ions emitted during the whole experiment had a value of M within the range 5·5 to 7·5. The values are immediately explicable on the assumption that the ions given off by Li_2So_4 are atoms of the metal which have lost an electron and that the heavier ions are due to some adventitious impurity. The atomic weights of lithium and potassium are 7·00 and 39·10 respectively, so that it is natural to attribute the heavier ions to salts of potassium. The fact that the potassium ions come off first is in agreement with the results of experiments on flames, which show that the conductivities produced by equivalent weights of salts of different alkali metals increase rapidly with the atomic weight of the metal. Thus while potassium salts are present one would expect the emission due to them to mask that due to the lithium salt, even though this might form the greater proportion of the salt dealt with.

These experiments and innumerable others, some of which will shortly be described, establish quite definitely the conclusion that *the positive ions emitted by heated salts are charged atoms of some metal.* This metal is not necessarily a constituent of the salt which appears to ·be under investigation, but may arise from the presence of some impurity which has a greater power of emitting positive ions.

The sulphates of the remaining alkali metals gave less evidence of the presence of impurities than did that of lithium. Thus the extreme variation of M for potassium sulphate was found to be from 35·5 to 37·0 during 60 hours of heating. None of the salts except lithium sulphate showed a double maximum in the magnetic field, although sodium sulphate and cæsium sulphate both gave somewhat exceptional values on first heating. This may be due to the presence of foreign ions in insufficient amount to give rise to a distinct maximum. The initial values for sodium were rather high, indicating the presence of potassium. The exceptional value for cæsium rests only on one observation, and it is also uncertain owing to the small deflexions given by the relatively heavy ions from this substance. The completeness of the separation of the maxima in the case of lithium is, of course, favoured by the large disparity of atomic weights when compared with the other pairs of metals, as well as by the relative amount of impurity already alluded to.

The final values of ϵ/m and of M which were deduced from the positions of the maxima characteristic of the basic metal of the salt under investigation are collected in the following table :—

Substance.	Time Heated (hours).	ϵ/m E.M. Units.	Actual M.	Average Value of M.	Atomic Weight.
Li_2SO_4	12	1760	5·5		
,,	44	1735	5·57	6·2	7·00
,,	52	1300	7·43		
Na_2SO_4	8	413	23·4		
,,	24	430	22·5	22·5	23·0
,,	13	439	22·0		
,,	15	439	22·0		
K_2SO_4	0	261	37·0		
,,	6	261	37·0		
,,	24	261	37·0		
,,	36	272	35·5	36·5	39·0
,,	42	266	36·3		
,,	60	266	36·3		
Rb_2SO_4	—	101	96	96	85·5
Cs_2SO_4	0	101·8	95		
,,	18	59·1	163	140	132·8
,,	23	59·1	163		

All the corresponding numbers in the last two columns differ by less than the possible experimental error, thus proving that the ions are atoms of the basic metal which have lost one electron. To be quite precise what is proved strictly is that the ions are made up of n atoms which have lost n electrons; but it is extremely unlikely that n is different from unity. To extend the proof tests were made with sodium fluoride and sodium iodide as well as sodium sulphate. In each case the value of M agreed with the atomic weight of sodium to within 5 per cent, which is about the accuracy claimed for the method used. Thus the acid constituent of the salt has no influence on the nature of the ions emitted by the salts of the alkali metals.

Some salts of the alkaline earth metals have been examined by the writer[1] and an exhaustive investigation of this group

[1] " Phil. Mag.," Vol. XXII, p. 669 (1911); Vol. XXVI, p. 452 (1913).

has been made by Davisson.[1] The measurements in each case were made by the slit method. One of the most interesting results of these experiments is that they afford no evidence of the existence of positive ions consisting of atoms of the basic metal which have lost two electrons; although since these metals are divalent the possible occurrence of such ions is indicated by electrolytic phenomena. For the sake of brevity we shall denote an ion consisting of an atom which has lost one electron by the symbol M_+ where M is the chemical symbol of the element. A divalent positive ion may be denoted in a similar manner by M_{++}. The measured values of the electric atomic weights demonstrate that positive ions having the constitution Ba_+ are given off when the following barium salts are heated: $BaSO_4$, $BaCl_2$ and BaF_2. Ions having the constitution Sr_+ have been shown to be emitted by the following salts of strontium: $SrSO_4$, $SrCl_2$, SrF_2, and SrI_2. In some of these cases there was evidence of the presence of ions having a value of M close to that for K_+. These were probably due to contamination of the preparations by salts of potassium. There is no mistaking the presence of the ions Sr_+ and Ba_+, as the values of M for them are very different from the values for K_+ and Na_+ which are the commonest adventitious impurities.

The case of calcium is not so clear, as the experiments are not accurate enough to distinguish between the values of M for Ca_+ (40·1) and K_+ (39·1). The same uncertainty arises in regard to magnesium where sodium may be an impurity. The respective values here are $Mg_+ = 24·3$ and $Na = 23·0$. A careful consideration of the conditions which govern the emission of the ions points to the conclusion that a considerable number of calcium salts emit Ca_+ and that some magnesium salts emit Mg_+. No evidence of the existence of Ca_{++} or Mg_{++} has been found. The salts of beryllium which have been examined have been found only to give ions with values of M corresponding to K_+ or Na_+ or to a mixture of these bodies.

The haloid salts of the metals of the zinc group furnish,

[1] " Phil. Mag.," Vol. XXIII, pp. 121, 139 (1912).

with one exception (p. 284), the only examples which have so far come to light of the existence of polyvalent positive ions. The theoretical value of M for Zn_+ is 65·4 and for Zn_{++} 32·7. Values of M close to 65 have been obtained[1] for the ions emitted from $ZnSO_4$ and $ZnCl_2$, indicating that the ions from these salts are singly charged atoms, as in the cases already considered. The mean value of M for the ions from $ZnBr_2$, on the other hand, has been found to be 50, and for ZnI_2, when first heated, four determinations by the quicker balance method gave values between 28·8 and 34·2. These

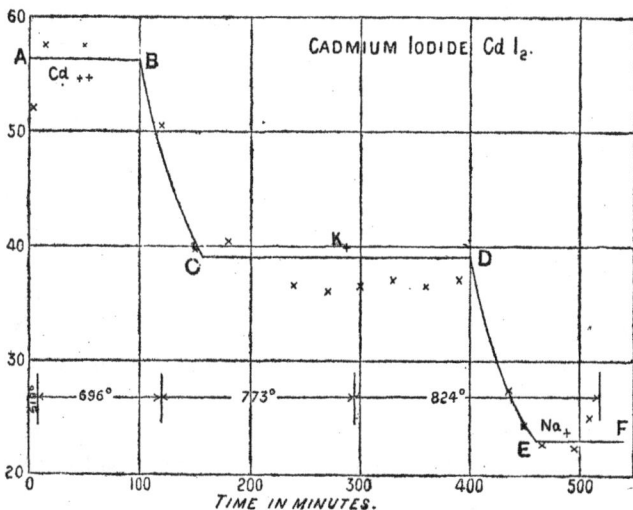

Fig. 32.

results point to the conclusion that the ions from the zinc haloids in general consist of a mixture of Zn_+ and Zn_{++} the proportion of Zn_{++} increasing with the atomic weight of the haloid constituent. The changes with the time which were noticed when a specimen of cadmium iodide was heated and tested by the balance method are shown in Fig. 32. The values of M are indicated on the vertical scale and the duration of heating on the horizontal scale. As the experiment progressed the emission at a constant temperature diminished; so that the temperature had to be raised from time to time in order to obtain a convenient current. The corresponding

[1] Richardson, " Phil. Mag.," Vol. XXVI, p. 452 (1913).

temperatures are also indicated in the figure. The straight lines AB, CD, and EF indicate the theoretical values of M for Cd_{++}, K, and Na_+. It will be seen that the experimental values, which are denoted thus X, jump successively from one of these lines to another. The demarcation between the ions characteristic of the salt and those due to impurities is not always so sharp as in this particular example.

Less work has been done with salts of metals of the other chemical groups. A test made with manganous chloride ($MnCl_2$) by the balance method gave an initial value M = 33·9. The value of M rose to a maximum of about 80 in 65 minutes and then fell to 39 at the end of 90 minutes. In this case it seems probable that a number of different kinds of positive ions are emitted in succession and that the numbers found are the average values of M for a mixture. The emission has not been found to be persistent enough to enable measurements to be made by the slit method so as to test the question of homogeneity as well as to determine the value of M.

A considerable number of salts have been tested and found to give evidence of the emission only of K_+ or Na_+ or a mixture of these ions, arising presumably from contamination with alkaline salts as impurities. Among these may be mentioned: ferric chloride (K_+), aluminium phosphate (Na_+ and K_+), barium phosphate (K_+), beryllium sulphate (K_+), and beryllium nitrate (Na_+ and K_+). The symbol in brackets indicates the nature of the ions as deduced from the experimental value of M.

A. T. Waterman[1] has investigated the following salts, viz. : AgCl, AgI, $PbCl_2$, $PbBr_2$, $PtCl_2$, AlF_3, Cu_2Cl_2, $CuCl_2$ and MoS_2 (molybdenite). Cu_2Cl_2 gave indications of Cu_{++} and the mineral molybdenite of Mo_+. Apart from these the only positive ions emitted by any of these salts appeared to have values of M corresponding to the impurities K_+ and Na_+.

The heavy negative ions emitted by certain haloid salts have been described in Chapter III, p. 107. The value of M indicates that these are atoms of the halogen constituent in combination with a single negative electron.

[1] "Phil. Mag.," Vol. XXXIII, p. 225 (1917).

The experiments under consideration have afforded no evidence of the existence, in the positive emission from salts, of any ions which are not charged metallic atoms, either of the basic element of the salt or of one of the alkali metals present as an impurity. We have seen in the previous section that the emission of positive ions from hot salts may be greatly increased at a constant temperature by the presence of a small quantity of various gases under certain conditions. The question arises as to whether the ions emitted in a dilute gaseous atmosphere under such conditions are still metallic atoms or whether atoms or molecules of the surrounding gas do not now carry part of the current. This question has been answered definitely in favour of the former alternative by some important experiments made by Davisson.[1] Using the slit method the value of M was measured for the ions given off by aluminium phosphate in hydrogen, air, and carbon dioxide, and by calcium sulphate in air. In each case experiments were made in a good vacuum and in the gases at various low pressures. The values of M were found to be independent of the nature and pressure of the gas until the pressures became so high that the collisions of the ions with the gas molecules caused serious deviations from the conditions required by the theory of the method of measurement. The limiting pressures were roughly: 0·05 mm. for CO_2, 0·12 mm. for air, and 1·4 mm. for hydrogen, in the case of the Na_+ ions given off by aluminium phosphate. The interference of the gas increased with its density, as was to be expected, and there was no evidence of the existence of any other effect of the gas on the value of M except that due to its mechanical interference with the motion of the ions. In accordance with these principles also the effect of air in the case of the Ca_+ ions from $CaSO_4$ was less than that observed with the Na_+ ions from aluminium phosphate, on account of the greater mass of the calcium ions.

The curves obtained with aluminium phosphate in hydrogen at a pressure of about 1 millimetre are shown in Fig. 33. The regularity and symmetry of these curves is strong

[1] Loc. cit., p. 145.

evidence as to the homogeneity of the ions. The arrows indicate the positions where maxima should appear for ions having values of $M = 1$ and $M = 2$ corresponding to H_+ and $(H_2)_+$ respectively. The absence of these maxima does not support the conclusion drawn by Garrett which was referred to at the beginning of this section. When salts are heated in a vacuum there is usually an appreciable emission of gas. The spectrum of this gas when an electrodeless discharge is passed through it has been examined by Horton[1] in the case of aluminium phosphate. All the lines except a few faint ones which were unidentified were found to be attributable to either Hg, H,

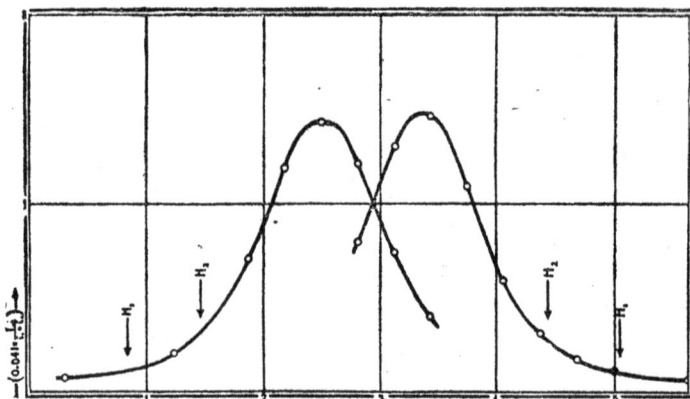

Fig. 33.

C, or O. A spectroscopic examination of the gases evolved by strontium chloride and strontium sulphate has been made by Davisson.[2] In each case only mercury and carbon monoxide lines were found. Davisson measured the value of M for the carriers of the positive emission simultaneously with the spectroscopic examination of the evolved gas. On first heating $SrSO_4$ the value of M was found to correspond to K_+. In all the other experiments the ions were found to be Sr_+. There was no indication of the existence of CO_+ in any of the experiments. In another set of experiments

---

[1] " Roy. Soc. Proc., A.," Vol. LXXXIV, p. 433 (1910).
[2] Loc. cit., p. 142.

Davisson measured the positive thermionic emission and the emission of gas from strontium chloride simultaneously. Both emissions varied in a fairly regular way with the time, but in an entirely independent manner, indicating that there was no direct connexion between them.

The measurements of e/m and of M which have been described in this section all refer to ions which are emitted from salts by the action of heat alone, and regarding whose emission the electric fields employed play only a secondary rôle. In addition it is necessary to mention that Gehrcke and Reichenheim [1] have measured the corresponding quantities for the anode rays from various salted anodes, whilst Knipp [2] has examined the nature of the ions in the canal rays from a Wehnelt cathode. In both these cases it seems likely that the ions are liberated as a result of the complicated electrical actions occurring and that the effects are of a character somewhat different from that of the phenomena contemplated in the rest of this book.

The Mobilities of Ions from Hot Salts.

Measurements of the mobilities of the ions, in air at atmospheric pressure, drawn away from the haloid salts of zinc heated to a temperature of about 360° C. have been made by Garrett and Willows.[3] The method used was one originally devised by McClelland.[4] The following values of the mobility in cms. per sec. per volt cm.$^{-1}$ were obtained for the positive ions: from $ZnCl_2$, 0·0062; from $ZnBr_2$, 0·0059; from ZnI_2, 0·0057. Somewhat variable values were obtained with the negative ions, the average for $ZnCl_2$ being about 0·02. The values for the positive ions are about 1/200 of the mobilities for X-ray ions in air at atmospheric pressure. This indicates that the ions from salts are more complicated structures. Since the measurements of e/m and of M have shown that the positive ions when originally emitted consist of atoms which have lost either one or two electrons, it follows that the

---

[1] " Phys. Zeits.," Jahrg. 8, p. 724 (1907).
[2] " Phil. Mag.," Vol. XXII, p. 926 (1911).
[3] *Ibid.*, Vol. VIII, p. 452 (1904).
[4] " Camb. Phil. Proc.," Vol. X, p. 241 (1899).

ions tested in these experiments must have become loaded up with uncharged matter. In all probability this consists for the most part of condensed salt vapours. It is to be remembered that when the gas is in the tube in which the mobilities are measured its temperature is much lower than when it was in contact with the hot salt. The values for the mobilities given above are similar to those found by McClelland[1] for the ions in gases drawn away from flames and incandescent metals.

A more complete investigation along similar lines has been made by Moreau,[2] using a large number of salts of the alkali metals. The supply of salt was obtained by bubbling air through solutions of various strengths. The air was then allowed to pass through a red-hot porcelain tube in which the salt became ionized. It then traversed the space between two co-axial cylindrical electrodes between which varying potential differences were maintained. When the stream of air flows at a known uniform rate a knowledge of the dimensions of the cylindrical electrodes and of the potential difference required to drive all the ions of given sign to the central electrode is sufficient to determine the mobility. The measurements of the mobility were made after the air had travelled various distances from the hot porcelain tube; so that the temperature of the ionized air had dropped to various values between 15° C. and 170° C. The temperature of the tube in which the salt was ionized was comparable with 1000° C. The concentrations of the salt vapours were taken to be proportional to the strengths of the solutions through which the air was made to bubble. In these experiments there is, of course, a great difference between the temperature at which the ions are formed and that at which the various measurements are made.

Under conditions similar to those indicated, Moreau examined several questions in addition to that of the dependence of the mobility of the ions upon various factors. These ques-

---

[1] Loc. cit.

[2] "Ann. de Chimie et Phys.," June, 1906; "Bull. de la Soc. Sci. et Méd. de l'Ouest," 15, No. 2 (1906); 15, No. 4 (1906).

tions include the relation between the quantity of ionization and the concentration and temperature of the salt vapour, and the rate of recombination of the ions at various low temperatures. Several of the most important conclusions drawn by Moreau from these experiments are enumerated in the following list :—

1. The number of positive ions formed is equal to the number of negative. [This conclusion involves the assumption that the charges of the ions of opposite sign are equal ; strictly speaking, the equality demonstrated by the experiments is between the total quantities of electricity of either sign liberated in the form of ions.]

2. Assuming that the concentration of the salt vapour is proportional to that of the solution through which the air bubbles, the experiments show that the number of positive or negative ions formed varies as the square root of the concentration of the salt vapour for a given salt at a given temperature.

3. The mobilities are different for the ions from different salts, but with the same salt the mobility of the negative ions is equal to that of the positive ions. The mobilities diminish rapidly as the temperature of the stream of air is reduced. With the salts of the alkali metals the mobility of the ions varies inversely as the cube root of the concentration of the salt vapour, for a given salt at a given temperature. These statements are illustrated by the values of the mobilities in cm. sec.$^{-1}$ per volt cm.$^{-1}$ given in the following table :—

Potassium Bromide.

Temperature °C.	170	110	100	70	30	15
N	0·42	0·20	0·15	0·09	0·046	0·012
N/4	0·72	0·35	0·27	0·16	0·046	0·026
N/16	0·95	0·57	0·35	0·28	0·083	0·026

Potassium Nitrate.

Temperature °C.	170	110	100	70	30	15
N	0·28	0·14	0·08	0·09	0·033	0·012
N/4	0·51	0·26	0·17	0·15	0·033	0·026
N/16	0·80	0·40	0·33	0·21	0·068	0·040

Rubidium Chloride.

Temperature °C.	170	—	100	70	30	15
N	0·73	—	0·30	0·19	0·084	0·021

The temperatures given are those of the air in the tube in which the measurements were made. The row of figures opposite the letter N gives the mobilities for the ions from solutions containing 1 gram molecule of the salt per litre, when the temperature of the ionized vapour has fallen to the temperature immediately above. The numbers opposite N/4 are for a solution of 1/4 of this strength, and so on.

All the values of the mobilities are much smaller than those for the ions in salted flames at a high temperature. The largest values are somewhat smaller than those for X-ray ions at atmospheric pressure, and the smallest are larger than those for the ions liberated during the oxidization of phosphorus. The structure of the ions is thus intermediate between those of these two classes. The variation of mobility with salt concentration is accounted for if one supposes that the ions are loaded by the condensation of salt vapour on primitive ions similar to those observed at very low pressures. We have seen already that the properties of the ions from the haloid salts of zinc agree with this hypothesis.

4. The coefficient a of recombination of the ions in the relatively cold tube at some distance from the source of ionization varies with temperature and salt concentration in the same way as the mobility of the ions. Its value in fact is in agreement with Langevin's formula

$$a = 4\pi(k_1 + k_2)\epsilon \qquad . \qquad . \qquad . \qquad (4)$$

when k_1 and k_2 are the respective mobilities of the positive and negative ions, here equal, and ϵ is a proper fraction which approaches unity at low temperatures.

5. The proportion of the salts which becomes ionized increases rapidly with the temperature of the hot tube. From the variation with temperature the energy required to ionize one gram molecule of the salts KCl, KBr, KI, and KNO_3 appears to be about 60,000 calories. This value is of the same order as that for the energy required to liberate the corresponding number of ions in other cases of thermionic emission.

J. J. Thomson [1] has shown that when ionization takes place

[1] " Conduction of Electricity through Gases," 2nd edition, p. 101 (Cambridge, 1906).

in a thin layer close to one of two parallel plates, and the currents are small compared with the saturation value, the mobility k of the ions may be obtained from the equation

$$k = \frac{32\pi}{9V^2} il^3 \qquad . \qquad . \qquad . \qquad . \qquad (5)$$

where V is the potential difference and l the distance between the plates, and i is the current. This method has been used by Garrett[1] to measure the mobilities of both positive and negative ions from a number of salts at comparatively low pressures. Some of the values of k thus obtained are given in the following tables :—

POSITIVE IONS IN AIR AT 215° C.

Pressure P (mms. of mercury).	ZnI_2.	BiI_3.	PbI_2.	CdI_2.	$k \times P.$ for BiI_3.
10	0·055	0·06	0·061	0·12	0·60
15	—	0·055	0·054	—	0·825
20	0·044	—	—	—	—
25	—	0·041	0·036	0·087	1·025
30	0·035	—	—	0·073	—
35	—	0·032	0·032	—	1·120
40	0·031	0·03	—	—	1·20
45	—	0·027	0·025	0·06	1·215
50	0·024	0·023	—	—	1·15
55	—	—	0·021	—	—
60	0·022	—	—	—	—
70	—	0·014	—	0·05	0·980
90	—	0·009	—	—	0·810

NEGATIVE IONS IN AIR AT 215° C.

Pressure P (mms. of mercury).	ZnI_2.	CaI_2.	BaI_2.	$P \times k$ for ZnI_2.
10	1·05	1·10	1·11	10·5
20	—	—	0·74	—
30	0·56	0·56	—	16·80
35	—	—	0·45	—
40	0·42	0·33	—	16·80
50	0·35	0·22	—	17·50
60	—	0·2	0·29	—
80	0·21	—	—	16·80

In these experiments the air in which the mobilities were measured was at the same temperature as the salt which caused the ionization. Unless the structure of the ions changes with the pressure of the gas the product of the mobility by the pressure should be independent of the pressure.[2] The last column in each table shows that this is approximately satisfied except at the lowest pressures. The value of the

[1] " Phil. Mag.," Vol. XIII, p. 739 (1907).
[2] Cf. Langevin, " Annales de Chim. et de Phys.," Vol. XXVII, p. 28 (1903) ; O. W. Richardson, " Phil. Mag.," Vol. X, p. 177 (1905).

product is also of the same order of magnitude as that given by the results of Garrett and Willows' experiments at atmospheric pressure in corresponding cases. This shows that the structure of the ions does not change much from atmospheric pressure down to the pressures used in these experiments. Garrett also found that the mobilities of the ions from CdI_2, ZnI_2, PbI_2, and BiI_3, at pressures between 10 and 25 mm.,

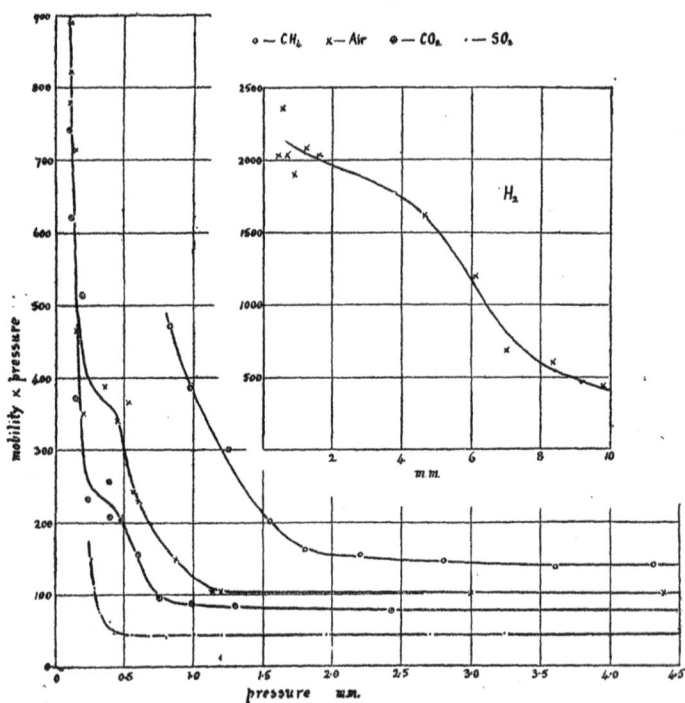

FIG. 34.

increased rapidly as the temperature was increased from 185 to 215° C.

Experiments on the positive ions from aluminium phosphate in H_2, CH_4, air, CO_2, and SO_2 have been carried down to much lower pressures by Todd,[1] who measured the mobilities by a modification of Langevin's[2] method. The values of the product mobility × pressure given by these experiments are

[1] " Phil. Mag.," Vol. XXII, p. 791 (1911).
[2] " Annales de Chim. et de Phys.," Vol. XXVIII, p. 289 (1903).

exhibited by the curves in Fig. 34. The value of the product is much larger than in the cases hitherto dealt with, probably owing to the high temperature of the salt. With each gas the product is practically constant at the higher pressures until a certain critical pressure is reached, beyond which any further reduction in the pressure causes an enormous increase in the product. At these pressures the complex structures present at higher pressures evidently begin to dissociate. The values of the product for the higher pressures were much the same as those for X-ray ions in the corresponding gases at atmospheric pressure. This agreement is probably fortuitous as the values of the product, at any rate in other similar cases, depend very much on the temperature of the gas and of the hot salt. In the case of hydrogen there is an indication of approach to a new constant value of $k \times P$ at the lowest pressures. Something of this sort is to be expected in all gases, since the experiments described in the last section have shown that the ions at these lower pressures are charged metallic atoms ; so that there can be no further simplification in structure after this stage is reached. When water vapour was present the values of the mobilities at low pressures were found by Todd to be abnormal.

GENERAL CONSIDERATIONS.

The measurements of the electric atomic weights of the positive ions emitted by heated salts abundantly prove that the chief process concerned in the liberation of the ions consists of a decomposition accompanied by the emission of positively charged metallic atoms. This decomposition may in different cases be that of the salt which forms the bulk of the specimen under examination or that of some intermediate body whose formation is accompanied by little or no emission, or it may be that of some other salt, or of an intermediate product arising from such salt, which is present as an impurity in the specimen. The fact that complex changes of the emission with time have usually been observed when salts are heated indicates that the intervention of an intermediate body is a very general feature of the phenomena. The only case in which there is no clear evidence of the existence of a time factor is that of cadmium iodide in the absence of water vapour which was studied by Kalandyk (p. 264). Even here it is not absolutely certain that the whole of the initial rise was due to the time necessary for the vapours to acquire a steady condition in the space between the electrodes by diffusion; but if this is admitted it would appear that this case affords the only example of the primary decomposition of the principal salt merely which has so far been observed.

The important, and often predominant, part played by impurities in the case of many salts has frequently been demonstrated. Among the most convincing cases that of lithium sulphate discussed on p. 279 and that of aluminium phosphate may be mentioned. Ordinary laboratory specimens of "pure" aluminium phosphate have been found to vary greatly in their emissive power. Thus Sir J. J. Thomson[1] found a specimen of this salt to be much more active than any of a large number of other salts which he tried, whilst a specimen examined by the writer[2] did not appear to be remarkable in this respect. In the belief that the activity of this substance is mainly attributable to the presence of alkaline impurities the writer[3] prepared a specimen of aluminium phosphate which one would expect to be comparatively free from such contamination by using only the materials aluminium chloride, phosphorus pentoxide, ammonia, and water, all of which had previously undergone distillation. As was expected this preparation gave a very small positive emission. After heating for a few minutes the emission was only about $\frac{1}{150}$th part of that from Kahlbaum's "pure" aluminium phosphate under similar conditions. This result has been confirmed by Horton.[4] By measurements of the electric atomic weights Davisson[5] has shown that the positive ions given off by the commercial "pure" aluminium phosphate

---

[1] "Camb. Phil. Proc.," Vol. XIV, p. 105 (1907).
[2] "Phil. Mag.," Vol. XXII, p. 698 (1911).   [3] Loc. cit.
[4] "Roy. Soc. Proc., A.," Vol. LXXXVIII, p. 117 (1913).
[5] "Phil. Mag.," Vol. XXIII, p. 144 (1912).

consisted at first of K_+. These were followed later by Na_+ which formed the bulk of the emission. The presence of sodium was also detected by the spectroscope whilst its disappearance when the emission has decayed to a small value has been demonstrated in the same way by Horton.[1] The emission from the specially pure aluminium phosphate after heating for a short time is so small that it is perhaps questionable whether it is not attributable to the underlying platinum, although Davisson[2] obtained indications of Al_+ by measuring the atomic weight of the carriers. Another case which illustrates the importance of looking out for the presence of alkaline contaminants is that of cadmium iodide considered on p. 283. Emissions which are conditioned mainly by the presence of impurities are apt to decay much more rapidly than those from substances like the involatile alkaline salts, where the rôle played by impurities is of a subordinate character. This is well shown by the following table of the currents from various substances, after heating for different times at 1190° C., given by Horton :—[3]

Substance.	At Start.	Positive Emission ($I = 10^{-8}$ Ampere).				
		1 Min.	2 Mins.	10 Mins.	50 Mins.	100 Mins.
Platinum	183	18·0	6·9	2·5	1·74	1·24
Pure aluminium phosphate	2040	201	87	18	5·6	3·6
Sodium phosphate . .	2550	3350	4430	5650	3750	1600
Sodium pyrophosphate .	3380	5220	5600	5270	2400	940
Impure aluminium phosphate . . .	7560	3220	1450	250	59	32

In the foregoing treatment the increase of positive emission with rising gas pressure which is observed when a number of salts are heated on strips of metal has been attributed to a mechanical action of the gas in interfering with the escape of a volatile ionizable product from the neighbourhood of the hot anode. In view of the complexity of the phenomena observed in gaseous atmospheres, and since this suggestion offers only a partial explanation of the observed facts, it seems desirable very briefly to consider some of the other views which have been put forward to account for this effect. Garrett[4] suggested that the increased currents might be due to the action of neutral

[1] " Phil. Mag.," Vol. XXII, p. 698 (1911).

[2] *Ibid.*, Vol. XXIII, p. 144 (1912).   [3] Loc. cit.

[4] " Phil. Mag.," Vol. XX, p. 573 (1910).

doublets shot out from the salts in ionizing the gas through which they passed. This view is subject, among other disadvantages, to one which is quite fatal. It fails to account for the fact that the increased current observed when the salt is positively charged is entirely absent when the salt is charged negatively. At one time Horton[1] held the view, based on the older values of the electric atomic weights for the positive ions from hot metals, on the identity of the kinetic energy of these ions with that of those emitted by aluminium phosphate, and on the detection of carbon and oxygen lines in the spectrum of the gas evolved by heated aluminium phosphate, that the ionization by salts consisted in the emission of carbon monoxide molecules in the positively charged condition. Such a position is, of course, untenable as a general account of the emission of these ions, in view of the various determinations of their electric atomic weights which have recently been recorded. Horton[2] has since modified it so as to make the gaseous ions carry only the additional current obtained on increasing the pressure, and has concluded that the effect is not a peculiar property of carbon monoxide but one which is common to the various gases, hydrogen, air, carbon monoxide, and carbon dioxide, which have been tested for it. The weak point of this position is that it does not account either for the absence of the effect when the salts are heated in a closed tube instead of on a strip, or for the experimental results obtained by Davisson which were described on p. 285. These results have been criticized by Horton[3] on the grounds that the experiments were made at temperatures so low that the number of gaseous ions would be expected to be inappreciable, and that, in any event, when the pressure is raised sufficiently for gaseous ionization to become effective the method of measurement fails owing to the interference of the gas molecules with the motion of the ions. The temperatures of the experiments are not stated at all clearly in Davisson's paper, but in many of them they were sufficiently high to

---

[1] " Roy. Soc. Proc., A.," Vol. LXXXIV, p. 433 (1910).

[2] *Ibid.*, Vol. LXXXVIII, p. 117 (1913).

[3] *Ibid.*

ensure, particularly in the case of hydrogen, the presence of a sufficiently large additional emission due to the gas at pressures so low that the method of measurement worked satisfactorily. Moreover, the deviations at high pressures in the apparent values of e/m are exactly such as would be expected from the mechanical interference of the gas molecules, and there is no indication of a specific influence of the gas on the nature of the emitted ions. No doubt the point is an extremely difficult one to decide with certainty, but it seems to the writer that the balance of evidence is definitely against the view that any considerable proportion even of the increased emission observed when salts are heated on strips of metal in a gaseous atmosphere at a low pressure is carried by charged atoms or molecules of gas. There is no doubt that the positive ions emitted by salts heated in a vacuum are not of this character. It is quite possible, and indeed rather likely, that when salts are heated in a gaseous atmosphere such ions are liberated at the hot electrode to some extent, but the experiments seem to show that if they exist they form an insignificant proportion of the total electrical emission. This part of the current would be expected to be most important at high pressures. Unfortunately it is only at low pressures that the nature of the original ions is discoverable by direct experiment.

The only experiments which have been made on the kinetic energy of the ions emitted by hot salts are some by F. C. Brown[1] who used ordinary " pure" aluminium phosphate. These show that the kinetic energy has the same value and mode of distribution as that of the ions emitted by hot metals.

A comparison of the results of this chapter with those described in the two chapters preceding shows that there is an exceedingly close parallelism between the emission of ions from salts and from freshly heated metal wires. This parallelism is not merely one which affects the more general phenomena which characterize the two emissions, such as the typical

---

[1] " Phil. Mag.," Vol. XVIII, p. 663 (1909).

relations between current and potential difference or current and temperature, and the charges and kinetic energies of the ions, but it is one which often extends in a very surprising way into the minute details of the two groups of phenomena. Thus the peculiar time changes when salts are first heated and the changes in the currents from salts due to a sudden alteration in the applied potential difference are very similar to effects which have often been observed with newly heated metals. In both cases, in the majority of instances, the ions emitted possess electric atomic weights corresponding to potassium or sodium. One might be tempted to infer from this that the effects exhibited by metals arise from alkaline saline impurities ; but such a conclusion cannot be considered to rest on a substantial foundation since the alkaline elements if dissolved or alloyed with the metals might give rise to effects similar to those caused by their salts. The positive emission from fresh wires is certainly not attributable to superficial saline impurities merely, since the most drastic treatment of the surface with acids, including hydrofluoric acid, fails to remove it. Perhaps the most noticeable difference between the emission from salts and that from fresh metals is the absence in the latter case of the response of current to change of pressure observed when salts are heated on metal strips. This would be expected if the alkaline atoms are completely ionized when emitted from metals, and, of course, the difference could be accounted for in a number of other ways. A surprising feature which the two groups of phenomena have in common is the way in which ions whose electric atomic weights correspond to K_+ and Na_+ turn up when the treatment would have led one to anticipate ions derived from one of the commoner gases.

So far as the relative efficiency of different salts in emitting ions when they are heated is concerned, it is clear that the degree of electropositiveness of the metallic constituent is a most important factor. The writer's experience is that the salts of the alkali metals are the leaders in this kind of activity, the comparative efficiency of the metals within this group increasing with increasing atomic weight. Superficially, at any rate, it would appear that volatility in salts is a factor conducive to ionization; at any rate, a number of volatile salts, including the haloid compounds of zinc and cadmium, are notable in this respect. There does not appear to be any very close connexion between the emission of positive ions from heated salts and the ionization of these salts in aqueous solution, as the solutions of the haloid salts of cadmium are distinguished for their relatively low electrical conductivity. Salts like aluminium phosphate which, though inactive if pure, generally emit ions owing to the presence of some impurity, often seem to give rise to effects larger than would be expected from the nature and amount of the impurity present. It may be that the activity of a given salt is increased, if equal quantities are compared, when a given amount of it is disseminated throughout a relatively large quantity of an inactive diluent. Such an effect would be similar to the effect of mixtures of salts in facilitating phosphorescence.

CHAPTER IX.

IONIZATION AND CHEMICAL ACTION.

THE difficulty, already alluded to, which frequently occurs when we attempt to discriminate between chemical and thermal action as the cause of ionization, suggests the propriety of closing this volume with a brief account of a number of interesting cases of gaseous ionization whose origin has generally, or at least frequently, been assigned to chemical action. Such phenomena have been known for over a century. Pouillet,[1] for example, observed that the air in the neighbourhood of a burning carbon rod acquired a positive charge whilst the rod itself became negatively charged. A jet of burning hydrogen was also found to be negatively charged, the surrounding air being positively charged. Similar effects with burning coal were recorded by Lavoisier and Laplace.[2] It seems likely that these effects are due to the high temperature of the materials rather than to the chemical actions taking place, although it is perhaps rash to hazard such an opinion in default of a more accurate investigation, such as the phenomena seem to merit.

Lavoisier and Laplace[3] also discovered that when iron is dissolved in sulphuric acid the hydrogen evolved contains a large excess of positive electrification. It has since been found that the gases liberated by chemical or electrolytic action from solutions almost invariably exhibit electrical conductivity. These effects have been investigated by Enwright,[4] Towns-

[1] "Pogg. Ann.," Vol. II, pp. 422, 426.
[2] "Phil. Trans.," 1782.
[3] "Mémoires de l'Académie des Sciences," 782.
[4] "Phil. Mag.," (5), Vol. XXIX, p. 56 (1890).

end,[1] Kosters,[2] H. A. Wilson,[3] Meissner,[4] Bloch,[5] Reboul,[6] and de Broglie and Brizard.[7] The ions present in these gases are bodies of considerable size. Thus Townsend found that their mobilities in an electric field were only about 10^{-4} of those of Roentgen ray ions in air under similar conditions as to pressure and temperature. When the gases liberated by electrolysis were passed into a vessel containing water vapour, a dense cloud was formed whose weight was proportional to the charge·in the gas. By measuring the rate of fall of such a cloud under gravity, together with a knowledge of the weight of the cloud and of the magnitude of the electric charge present in it, Townsend was able to determine the charge e of these ions. The value found was $5 \cdot 1 \times 10^{-10}$ E.S.U. in excellent agreement with the value found subsequently for the corresponding quantity for gaseous ions from other sources.

In all these cases of the presence of ions in the gases liberated by the electrolysis of liquids and by chemical actions in the wet way, the effects are undoubtedly complicated by the occurrence of ionization due to bubbling and splashing. In fact, the recent experiments of Bloch[8] and of de Broglie and Brizard[9] seem to show that the ions arise entirely from the action of the bubbles of gas in bursting through the surface of the liquid. Thus Bloch found that if the surface of the liquid were covered with a layer of benzene or of a number of other liquids, the conductivity of the liberated gas disappeared completely. Further information about the experiments which have been made on the electrification caused by bubbling and splashing may be found in J. J. Thomson's "Conduction of Electricity through Gases," 2nd edition, page 427, and in the following papers :—

[1] "Camb. Phil. Proc.," Vol. IX, p. 345 (1897); "Phil. Mag.," (5), Vol. XLV, p. 125 (1898); "Camb. Phil. Proc.," Vol. X, p. 52 (1899).

[2] "Wied. Ann.," Vol. LXIX, p. 12 (1899).

[3] "Phil. Mag.," (5), Vol. XLV, p. 454 (1898).

[4] "Jahresber. für Chemie," 1863, p. 126.

[5] "Ann. de Chimie et de Phys.," (8), Vol. IV, p. 25 (1905); "C.R.," Vol CXLIX, p. 278 (1909); *ibid.*, Vol. CL, pp. 694 and 969 (1910).

[6] *Ibid.*, Vol. CXLIX, p. 110 (1909); Vol. CLII, p. 1660 (1911).

[7] *Ibid.*, Vol. CXLIX, p. 924 (1909); Vol. CL, p. 916 (1910); Vol. CLII, p. 136 (1911).

[8] *Ibid.*, Vol. CL, p. 694 (1910).   [9] *Ibid.*, p. 969 (1910).

Lenard, "Wied. Ann.," Vol. XLVI, p. 584 (1892); Lord Kelvin, "Roy. Soc. Proc.," Vol. LVII, p. 335 (1894); J. J. Thomson, "Phil. Mag.," (5), Vol. XXXVII, p. 341 (1894); Lord Kelvin, Maclean and Galt, "Phil. Trans., A." (1898); Kosters, "Wied. Ann.," Vol. XLIX, p. 12 (1899); Kaehler, "Ann. der Phys.," Vol. XII, p. 1119 (1903); Aselmann, "Ann. der Phys.," Vol. XIX, p. 960 (1906).

Air which has been drawn over phosphorus is capable of discharging both positively and negatively electrified conductors. This was first noticed by Matteuci.[1] The phenomenon has since been investigated by Naccari,[2] Elster and Geitel,[3] Shelford Bidwell,[4] Barus,[5] Schmidt,[6] Harms,[7] Goekel[8] and Bloch.[9] Barus noticed that the phosphorized air very readily formed clouds in a moist atmosphere. Both Bloch and Harms found that the currents through phosphorized air could be saturated if sufficiently large electromotive forces were applied. Bloch determined the mobility and the coefficient of recombination of the ions and found that both these quantities were only about one-thousandth part of the corresponding quantities for X-ray ions. The ions in phosphorized air are thus comparatively large structures and are probably loaded up with the compounds of phosphorus formed during the reaction. Both Bloch and Harms found that the number of ions formed was small compared with the number of molecules of oxygen which combined with the phosphorus. Barus showed that no ions are formed when chemically inactive gases such as hydrogen are passed over phosphorus, and Elster and Geitel showed that the ionization which occurs when air is passed over heated sulphur is small compared with that which arises during the slow oxidation of phosphorus. These experiments show that the ionization of phosphorized air is intimately

---

[1] "Encyclopædia Britt.," Vol. VIII, p. 622 (1855 edition).
[2] "Atti della Scienzi de Torino," Vol. XXV, p. 252 (1890).
[3] "Wied. Ann.," Vol. XXXIX, p. 321 (1890).
[4] "Nature," Vol. XLIX, p. 21 (1893).
[5] "Experiments with Ionized Air," by C. Barus (Washington, 1901).
[6] "Ann. der Phys.," Vol. X, p. 704 (1903).
[7] "Physik. Zeits.," 3 Jahrg., p. 111 (1902).
[8] *Ibid.*, 4 Jahrg. (1903).
[9] "Ann. de Chimie et de Physique," (8), Vol. IV., p. 25 (1905).

connected with the chemical action and is probably directly caused by it. However, the slow oxidation of phosphorus is exceptional when compared with most chemical reactions at low temperatures, inasmuch as it is accompanied by the emission of light. It seems probable that both the ionization and the emission of light are direct and simultaneous consequences of the chemical reaction, but the possibility that the ionization is an indirect photoelectric effect due to the action of the light emitted does not seem to be altogether excluded.

Another case of ionization, apparently caused by chemical action, in which phosphorus takes part has been observed by the writer.[1] At about 600° C. platinum reacts energetically with phosphorus vapour. During the occurrence of the reaction the platinum emits positive but not negative ions. After platinum has been left cold in contact with phosphorus vapour, a vigorous emission of positive ions takes place when the metal is heated subsequently. This decays at constant temperature like the positive emission from new wires. Overheating the wire was found to reduce the emission at the previous temperature temporarily. There was some recovery at constant temperature from the reduction due to overheating which was subsequently followed by the general decay at constant temperature already referred to. These phenomena suggest that the emission involves two distinct processes whose rates are altered to different extents when the temperature is changed. Increasing the temperature appears to reduce the quantity of the substance which immediately gives rise to the emission of the ions without destroying the parent substance to an equal extent. Similar changes due to sudden disturbances of the temperature have been found to characterize the emission of ions from heated salts (see p. 266). In some cases, though not invariably, the effect shown by salts is in the opposite sense to that just referred to. Thus when sodium phosphate or sodium sulphate was overheated, the emission at the original lower temperature was found temporarily to be increased.

[1] O. W. Richardson, "Phil. Mag.," (6), Vol. IX, p. 407 (1905).

The ionization, discovered by Le Bon,[1] which accompanies the hydration and dehydration of certain crystals has frequently been attributed to chemical action. The case which has attracted most attention is that of quinine sulphate. This substance, when allowed to cool after heating to a certain high temperature, phosphoresces and causes the surrounding gas to become conducting. Miss Gates[2] showed that the ionization was not caused by rays capable of penetrating the thinnest aluminium foil. She also found that the current from the salt was greater when it was positively than when it was negatively charged and that the hydration of a given amount of salt caused a greater conductivity than the dehydration. These results were confirmed by Kalaehne,[3] who concluded, in addition, that the hydration of a given amount of the salt at a fixed temperature liberated a constant quantity of electricity independently of the rate of hydration, although the actual instantaneous currents depend very considerably on the rate of hydration. Recent experiments by de Broglie and Brizard[4] suggest that in all these cases the ionization is only an indirect effect of the absorption or liberation of water vapour. Both the ionization and the luminosity observed with the sulphates of quinine and cinchonine seem to be due to minute sparks arising from the triboluminescence of the crystals of these substances which takes place during hydration and dehydration. Although it is almost impossible to saturate the currents from these substances the ions have a high coefficient of re-combination. Both the ionization and the scintillations increase as the pressure is reduced from atmospheric.

We have seen (p. 219) that the evidence is quite conclusive that the large emission of positive ions from freshly heated wires is not directly caused by chemical action between the hot wires and surrounding gases. This is clear since the effects are shown as well in a good vacuum as in a gaseous atmosphere and by platinum when surrounded by gases to

[1] "C.R.," Vol. CXXX, p. 891 (1900).
[2] "Phys. Rev.," Vol. XVIII, p. 135 (1904).
[3] "Ann. der Phys.," Vol. XVIII, p. 450 (1905).
[4] "C.R.," Vol. CLII, p. 136 (1911).

which it is believed to be chemically indifferent as by other metals when heated in various gases with which they react energetically. Indeed, Strutt's experiments (p. 205) led him to the conclusion that oxidation and reduction by gases were unfavourable to the emission rather than otherwise. These facts do not preclude the hypothesis that this emission is a direct consequence of chemical reactions affecting alkaline contaminants present in the metals. In fact, the changes in the emission with time at constant temperature and after sudden changes in such factors as the temperature and voltage which govern the equilibrium conditions definitely suggest that this emission is affected, directly or indirectly, by chemical changes. On the other hand, everything points to such changes being of an obscure character and nothing is definitely known either as to what the chemical changes are or as to the way in which they affect the emission.

The results referred to do not prove that an emission of ions may not occur as a consequence of chemical action between metals and surrounding gases. They only show that effects of this kind, such as have so far been looked for, are small in comparison with the large positive emission from new wires. There is, in fact, quite definite experimental evidence which, on a superficial examination at any rate, suggests that the emission of ions can occur as a direct result of chemical action between metals and surrounding gases. Thus Campetti[1] found an emission of positive ions when copper combines with oxygen or chlorine. In a later paper[2] he determined the mobilities of these ions and found they were relatively heavy bodies, probably composed of copper oxide. The emission during the oxidation of copper has been confirmed by Klemensiewicz,[3] who showed that it was small compared with the initial effect. He also found a similar effect when oxidized copper was reduced in hydrogen. The oxidation and subsequent reduction of both tungsten and iron wires were also examined. Tungsten gave larger

[1] "Atti di Torino," Vol. XLII; "N. Cim.," (5), Vol. XIII, p. 183 (1907).
[2] "Atti di Torino," Vol. XLVI, p. 180 (1911).
[3] "Ann. der Physik," (4), Vol. XXXVI, p. 805 (1911).

and iron much smaller effects than copper. Klemensiewicz also investigated the reversible oxidation and deoxidation of palladium and iridium, but was unable to detect the emission of either positive or negative ions. It is necessary to be rather cautious in the interpretation of the results of these experiments. In the writer's opinion it is not certain that the obvious conclusion that the emission is a direct consequence of the chemical action is the correct one. We have seen that it is often quite difficult to get rid of the last traces of the "initial emission," and even when this appears to have been accomplished what seem to be quite trivial changes in the conditions of the experiment will frequently revive the emitting substance to a considerable extent (see p. 202). Thus to make sure that the effects under discussion are really direct chemical effects it is at least necessary to make sure that the emission can be repeatedly obtained from the same material without diminution, under given conditions. So far this test does not seem to have been made.

We have seen (p. 239) that the steady positive emission from platinum in an atmosphere of various gases undergoes a temporary increase when the composition of the gas is changed. If this increase were confined to cases in which the gases interchanged had considerable chemical affinity for each other, like oxygen and hydrogen, one would be tempted to attribute it to a reaction between the gases in the surface layers of the metal. Such a view seems impossible, however, when we recollect that this effect has been observed when the chemically inert gas helium is introduced.[1] It would be interesting to see if the increase occurred if pure helium were replaced by pure argon or vice versa.

The part played by chemical action in connection with the ionization from hot salts has already been considered at some length in Chapter VIII, and there does not seem to be anything to add to the discussion there given. In fact, it is difficult to advance beyond the generalities already discussed, since there is no definite information either as to what the chemical reactions in question are, or as to how they affect

---

[1] Richardson, "Phil. Trans., A.," Vol. CCVII, p. 1 (1906).

the ionization. The action of hydrogen on $BaSO_4$ considered on p. 275 appears to a certain extent to furnish an exception to this statement.

EMISSION OF ELECTRONS UNDER THE INFLUENCE OF CHEMICAL ACTION.

The effects considered so far are noteworthy in two respects. In the first place, the emitted ions are of atomic or greater magnitude. There is no evidence of the emission of *electrons* as a result of any of these actions. In fact, in nearly every case in which the chemical origin of the ionization is not extremely doubtful, only positive ions are emitted. In the second place, we have seen that it is very questionable whether many of these affects can be considered a direct result of chemical action at all. The cases in which the connexion between ionization and chemical action appears to be most intimate are furnished by the oxidation of phosphorus, the action of phosphorus on platinum and the emission of ions from heated salts. Certain types of chemical action resulting in the liberation of electrons were first examined by Reboul [1] who investigated the following reactions: the oxidation of amalgamated aluminium and of sodium and potassium by moist air, the action of H_2S on silver and on the alkali metals, the action of CO_2 on the alkalies and that of nitrous fumes on copper. In all these cases ionization was observed when the various reagents were attacked at the ordinary laboratory temperature by the gases referred to, and in most cases more negative than positive ions were apparently emitted. In some of these cases it is doubtful whether ionization occurs unless the reaction is allowed to proceed with sufficient vigour to raise the temperature considerably, or until it has gone on long enough to form a layer of the products of the reaction over the surface of the liquid or solid reagent. Thus some of the effects observed are probably to be attributed to thermal emission or to electrical effects arising from bubbling in, or fracture of, the layer in question.

---

[1] " C.R.," Vol. CXLIX, p. 110 (1909) ; Vol. CLII, p. 1660 (1911).

Experiments which are not open to these objections, or at least not obviously open to them, have been made by Haber and Just.[1] These authors investigated the action of one or more of the following gases or vapours, viz. : H_2O, $COCl_2$, $CSCl_2$, HCl, O_2, Cl_2, Br_2, and I_2 on various dilute amalgams of the alkali metals, on cæsium, and on the liquid alloy of sodium and potassium. The experiments were made by allowing a fine stream or jet of the liquid reagent to flow into a dilute atmosphere of the gas. The current was then measured which passed from the jet to a surrounding cylindrical electrode under various conditions. Many of the experiments were made with the atmosphere of gas or vapour at a very low pressure and the jet or stream of drops was made to flow so fast that no observable tarnishing of the metal surface could be detected. The thickness of the layer of salt formed on one of the drops in an atmosphere of bromine in which an energetic electrical emission occurred is estimated by the authors as 3×10^{-7} cm. It was thus not more than a few molecules thick.

All the reactions referred to caused an emission of negative electricity from the metal but there was no positive emission when the reactions occurred at room temperatures. Experiments with the alloy of sodium and potassium at low pressures showed that the current was stopped by a magnetic field ; thus proving that the carriers of the discharge when first liberated from the reacting metal are electrons. This effect of a magnetic field disappears at higher pressures, probably owing to the electrons combining with the molecules of the gas. No emission from a jet of sodium potassium alloy could be detected in an atmosphere of hydrogen or of nitrogen. This result is not in agreement with a previous observation by J. J. Thomson[2] who found an emission of electrons when hydrogen was admitted to the alloy of sodium and potassium. The discrepancy could be reconciled if the hydrogen used by Thomson were not entirely free from moisture. In many of

[1] "Ann. der Phys.," Vol. XXX, p. 411 (1909) ; *ibid.*, Vol. XXXVI, p. 308 (1911); "Zeits. f. Elektrochemie," Vol. XVI, p. 275 (1910).
[2] "Phil. Mag.," (6), Vol. X, p. 584 (1905).

these cases the number of electrons emitted is very considerable in proportion to the amount of chemical action occurring. In a particular case in which the alloy of sodium and potassium was attacked by carbonyl chloride Haber and Just estimate that one electron was emitted for every 1600 molecules of salt formed, approximately. The negative ions given off by the amalgams of the alkali metals are apparently not electrons, as the currents from these bodies were unaffected by a magnetic field. In some of the reactions the electrons were found to be emitted with sufficient kinetic energy to charge up the neighbouring silver electrode even when they had to travel against a small opposing electric field. Thus with $COCl_2$ and NaK alloy, the silver plate was found to charge up by amounts varying from 0·7 to 1·2 volts negative to the alloy. When iodine reacted with the alloy, however, at least 1·3 volts accelerating potential difference was necessary to obtain an appreciable current. In the case of cæsium the corresponding energies were somewhat greater in every case. Thus with $COCl_2$ the electrode charged up to − 1·6 volts, with Br_2 to − 1·0 volt, and with iodine an accelerating potential difference of only 0·4 volt was necessary to detect the emission.

The maximum potentials obtained in the way just indicated do not, unfortunately, enable us to deduce the maximum kinetic energies with which the electrons are liberated by the chemical action, as no attempt has been made in the experiments referred to to correct for the effect of the contact difference of potential between the emitting metal and the receiving electrode. This contact difference of potential causes an electric field which affects the motion of the electrons in the space between the electrodes, but it does not affect the instruments used to measure the potential difference between the electrodes during the experiments. The allowance to be made for this is uncertain. Sodium potassium alloy is several volts electropositive to clean silver, and if this full potential difference were operative possibly as much as 3 volts would have to be added to the numbers given above. On this assumption the maximum emission energies in equivalent volts would be as follows: For NaK alloy and $COCl_2$ from 3·7 to 4·2, NaK alloy and Br_2 about 3·0, NaK alloy and I_2 1·7, Cs and $COCl_2$ 4·6, Cs and Br_2 4·0, and Cs and I_2 2·6. On the other hand, if the silver had been splashed with the alloy, a not impossible contingency in experiments of this character, the effective contact potential difference might lie anywhere between the full value and zero. Thus all it seems legitimate to assume from these numbers is that the maximum kinetic energy of the electrons liberated by the action of $COCl_2$ on sodium potassium alloy lies between the limits 0·7 and 4·2 equivalent volts with a similar possible range of uncertainty in the values for the other reactions.

Measurements of maximum potentials in this way are also open to another objection. If the emission velocities extend from zero to infinity, as in the case of thermionic electrons, such limiting potentials have no definite meaning as they indicate merely the stage at which the electron emission current is so attenuated by the retarding effect of the opposing voltage that it just compensates the current due to the various leakage effects which can never be entirely got rid of. If, however, there were a definite limiting velocity as in the photoelectric case this objection would not apply.

In the last few years[1] I have devoted a considerable amount of attention to the experimental investigation of these phenomena and have, to a certain extent, succeeded in reducing them to a quantitive basis. Most of the experiments were made with a liquid alloy of sodium and potassium having the composition NaK_2. The gases used, arranged in what appears to be the order of their activity, were chlorine (Cl_2), carbonyl chloride ($COCl_2$), water vapour (H_2O), hydrochloric acid (HCl) and air. No emission of electrons was observed in dry air. The liquid alloy was allowed to flow down a vertical tube and fell in a stream of small drops about 3 mms. in diameter from a minute orifice at the centre of a surrounding spherical or cylindrical metal electrode. These were suitably mounted in an exhausted glass vessel which was filled with the active gas at a low pressure. It was not possible to measure the pressure of the active gas but it was probably less

---

[1] "Phil. Trans., A.," Vol. CCXXII, p. 1 (1921).

than 0·001 mm. in all cases. The surrounding electrode was connected to the earthed pair of quadrants of a delicate electrometer with suitable capacity added when necessary. By this means the electron currents flowing were measured when various accelerating and retarding potentials were applied between the drops of alloy and the surrounding electrode. The result of a typical experiment, in which the cylindrical electrode and $COCl_2$ were used, is shown in Fig. 35. In every case it was found that the currents were practically constant for all accelerating potentials greater than two or three volts. As a matter of fact Fig. 35 shows a small diminution with rising

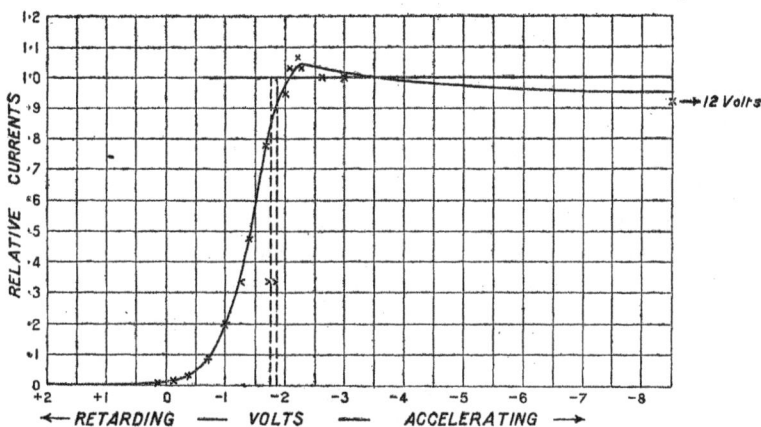

FIG. 35.

potential in this region but other tests showed a small increase and still others no change; so that too much significance should not be attached to small changes of this order. On the other hand, with apparent accelerating volts less than about two the currents decreased rapidly and approached the voltage axis gradually at small retarding voltages. Similar current-voltage curves were obtained with chlorine and water vapour but with hydrochloric acid the effects were too small to make any satisfactory measurements.

It is to be remembered that the voltages shown in Fig. 35 are those measured on a voltmeter in parallel with the gap between the drops and the surrounding electrode. These will not be the true operative potential differences on account of the considerable unknown contact difference of potential which is not recorded by the voltmeter. This contact difference of potential has been determined by a photoelectric method, and thus the position of the true zero on the voltage axis found, at the same time and under the same conditions as the chemical currents were measured. The writer and K. T. Compton[1] have shown that if ν_0 is the least frequency of light which will excite photoelectric emission from any given metal surface, if V_1 is the voltage at which the current-voltage curve with monochromatic light of frequency ν intersects the voltage axis, then the true zero on the voltage axis lies at the value V where

$$e(V - V_1) = 300 \cdot h(\nu - \nu_0).$$

It is thus necessary to determine ν_0 and the value of V_1 for some particular value of ν in order to find V. This has been done by well-known methods and the result shows that the true zero lies between the limits set by the two broken vertical lines which are 0·1 volt apart. If we take the true zero to lie anywhere within this region we see that, broadly speaking, and disregarding a little abnormality in the neighbourhood of zero volts, the chemical currents are saturated for all accelerating voltages and show a rapid diminution for all retarding voltages. It shares these features in common with small photoelectric currents and with small thermionic currents when the disturbing effects due to the heating current are eliminated. Pending further investigation it seems reasonable to attribute the small abnormalities in the chemical current in the neighbourhood of zero volts to the operation of the self-repulsion of the electrons and to the other causes which have been found to give rise to similar effects in this region in the thermionic case.

The manner in which the current-voltage curves approach the voltage axis on the retarding voltage side has been carefully examined and the evidence is quite definite that this approach is gradual as in the thermionic and not sharp as in the photoelectric case.

---

[1] " Phil. Mag.," Vol. XXIV, p. 575 (1912).

The shapes of the curves in the region of retarding voltage enables the distribution of energy among the emitted electrons to be ascertained. The number of emitted electrons having kinetic energies between u and $u + du$ is found, within the limits of experimental error, to be of the form

$$\frac{A}{k^2 T^2} e^{-\frac{u}{kT}} u\, du,$$

where A and (kT) are constants. If k is Boltzmann's constant this is a Maxwell distribution for the molecules of a monatomic gas at temperature T. For $COCl_2$ T is found to be $3300°$ K. and for chlorine (less accurately) $4900°$ K. The kinetic energy of the emitted electrons is thus the same as if they were in thermal equilibrium at the high temperature T. We have seen that recent results (p. 172) indicate that when electrons are emitted thermionically from a metal they sometimes appear to have a distribution of energy which would agree with a Maxwell distribution for some temperature which is higher than that of the metal. In spite of this I think it would be unwise to conclude *either* that thermionic emission is caused by some very elusive kind of chemical action *or* that the effects caused by these active gases on the alkali metals are in reality secondary thermionic effects caused indirectly by the local rise of temperature due to the chemical action. The evidence against the former view is overwhelming and that against the latter is very strong. It appears from the experimental data, for example, that the chemical saturation currents, which obviously are a measure of the rate of the chemical reaction, can be increased 800 fold without any noticeable change in the curves which express the energy distribution. It is inconceivable that an increase of some hundred-fold in the rate of chemical action would not affect profoundly the temperature attained. Various other conditions can be changed, such as the size and rate of formation of the drops, without noticeably affecting the shapes of the velocity distribution curves. It appears that the distribution is a fundamental property of the chemical reaction and not due to secondary phenomena.

If this similarity in the distributions is substantiated on a closer and more detailed examination it seems likely that it will be attributable to the common operation of the fundamental laws which govern both thermionic and chemical phenomena, such as may be termed quantum dynamics. From such a broad standpoint it is difficult to distinguish[1] between thermionic, photoelectric, photochemical, and chemical actions. It is important to remember that the reaction here under investigation is an irreversible one and it appears that even in such a case the products of the reaction have an energy distribution appropriate to that of thermal equilibrium at some particular temperature. This result is, of course, only established to the degree of accuracy of the experiments.

IONIZATION OF GASES BY HEAT.

There is no satisfactory a priori reason for expecting the emission of ions at a high temperature to be confined to matter in the solid and liquid states. It is, however, to be anticipated that the thermal ionization of gases will only be appreciable at the very highest temperatures, on account of the large value of the ionization energy of gases. This quantity, which has been measured by experiments on impact ionization and on photoelectric action, has in all cases been found to be much greater than the energy changes governing the liberation of an ion in the phenomena which have been considered in this book. Up to the present there is no evidence that purely thermal ionization has been observed in any of the commoner gases. It seems likely that the ions present in flames are to be attributed to the chemical actions occurring rather than to the direct effect on the gases of the high temperature which prevails. In the case of gases which have been heated in the presence of metal electrodes there is no certain evidence of the formation of ions except by interaction between the gases and the electrodes or by emission from the electrodes themselves.

[1] O. W. Richardson, " Phil. Mag.," Vol. XXVII, p. 476 (1914).

A possible exception to these statements is furnished by some experiments made by J. J. Thomson [1] on sodium vapour. He found that when a current was made to pass between two electrodes immersed in this vapour at about 300° C. metallic sodium collected on the negative but not on the positive electrode, indicating that sodium atoms in the vapour had dissociated into an electron and a positive sodium ion. The phenomenon could also be accounted for if the bombardment of the positive electrode by electrons present made it hotter than the negative electrode. The optical properties of sodium vapour make it probable that it will dissociate, in the manner indicated, below 1000° C. It is necessary to add that Thomson's experiments have been repeated by Fredenhagen [2] without success; and Dunoyer,[3] who has made very extensive experiments on this point, concludes that the conductivity of sodium vapour below 400° C. is of the same order of magnitude as that of the spontaneous ionization of gases at ordinary temperatures. Campetti,[4] however, finds that a marked ionization of sodium vapour sets in at a little above 400° C.

It is necessary also to make an exception in favour of salt vapours. In the case of cadmium iodide the evidence of the occurrence of ionization of the vapour is quite definite (p. 255); but, even in this case, the possibility that it arises by interaction with the electrodes is not absolutely excluded. In any event the phenomena in salt vapours are probably complicated by secondary chemical actions.

A theory of the ionization of gases by heat following these general lines has recently been elaborated by M. N. Saha [5] who has shown that it receives considerable support from spectroscopic phenomena in the chromosphere and from other astrophysical data.

[1] " Phil. Mag.," Vol. X, p. 584 (1905).
[2] " Phys. Zeits.," Vol. XII, p. 398 (1911).
[3] " Ann. de Chimie et de Phys.," Vol. XXVII, p. 494 (1912).
[4] "Atti di Torino," Vol. LIII, pp. 519, 608 (1918).
[5] " Phil. Mag.," Vol. XL, pp. 472, 809 (1920).

INDEX OF NAMES.

SUBJECT INDEX.

PRINTED IN GREAT BRITAIN BY THE UNIVERSITY PRESS, ABERDEEN